Blunt Instrument

Blunt Instrument

Why Economic Theory Can't Get Any Better . . . Why We Need It Anyway

Alex Rosenberg

The MIT Press
Cambridge, Massachusetts
London, England

© 2025 Massachusetts Institute of Technology

All rights reserved. No part of this book may be used to train artificial intelligence systems or reproduced in any form by any electronic or mechanical means (including photocopying, recording, or information storage and retrieval) without permission in writing from the publisher.

The MIT Press would like to thank the anonymous peer reviewers who provided comments on drafts of this book. The generous work of academic experts is essential for establishing the authority and quality of our publications. We acknowledge with gratitude the contributions of these otherwise uncredited readers.

This book was set in Stone Serif and Stone Sans by Westchester Publishing Services. Printed and bound in the United States of America.

Library of Congress Cataloging-in-Publication Data

Names: Rosenberg, Alexander, 1946– author.
Title: Blunt instrument : why economic theory can't get any
    better . . . why we need it anyway / Alex Rosenberg.
Description: Cambridge, Massachusetts : The MIT Press, [2025] | Includes
    bibliographical references and index.
Identifiers: LCCN 2024019132 (print) | LCCN 2024019133 (ebook) |
    ISBN 9780262049658 (hardcover) | ISBN 9780262382366 (pdf) |
    ISBN 9780262382373 (epub)
Subjects: LCSH: Economics—Philosophy. | Economics—History.
Classification: LCC HB71 .R668 2025 (print) | LCC HB71 (ebook) |
    DDC 330.01—dc23/eng/20240815
LC record available at https://lccn.loc.gov/2024019132
LC ebook record available at https://lccn.loc.gov/2024019133

10 9 8 7 6 5 4 3 2 1

For Christopher and Alex
With thanks for making it fun

Contents

1 Why We Need to Understand Economic Theory   1
2 Economics Won't Budge, Can't Budge, from Rational Choice Theory   13
3 Adam Smith Is Still Setting the Agenda for Economic Theory   39
4 Who Invited Calculus?   59
5 Money—Who Needs It? Not Economic Theory!   73
6 Back to the Future with New Classical Macroeconomics   99
7 Economic Theory's War against Profit   133
8 It's All Game Theory Now   159
9 Fated to Fail: The Financial Markets   181
10 Economic Theory as Institution Design   197

Notes   227
Bibliography   233
Index   237

1 Why We Need to Understand Economic Theory

Many people would like to know what economics is really all about, and especially to understand what's going on in economic theory. We're pretty much in the same boat as those who want to learn what quantum theory means without having to solve the problems at the end of each chapter in a physics textbook. There may even be more people who are curious about economics without wanting to become economists themselves. There certainly should be more people curious about economic theory than quantum theory or the general theory of relativity. After all, the economy matters to us every day, but gravity waves and the Higgs boson are less noticeable influences on the quality of our lives.

This book is an outsider's guide to what is going on in economic theory. It's an area we all need to know about, not just because the economy affects us all the time, but also because economists—the users of economic theory—help shape for good and ill the ways the economy affects us. Economists guide government and business from day to day, month to month, year to year, making our lives better or worse. We and the leaders who allow themselves to be guided by economists need to understand whether and to what extent we should do so.

No less an economist than John Maynard Keynes expressed this point in a memorable quotation: "The ideas of economists and political philosophers, both when they are right and when they are wrong, are more powerful than is commonly understood. Indeed the world is ruled by little else. Practical men, who believe themselves to be quite exempt from any intellectual influence, are usually the slaves of some defunct economist."[1] And not just practical businesspeople and government—the rest of us are, if not slaves, then at least sometimes victims and (if we're lucky) beneficiaries of both defunct and up-to-date economists. Maybe if we learn a little about the tools economists employ—their strengths and their limits—we'll be in a better position to evaluate economists' advice. After all, it hasn't always been very good. In fact sometimes it has been catastrophic. Everyone but a Gen Zer is old enough to remember how bad economic guidance was in the years before and after the subprime mortgage crisis.

The trouble is that there are barriers to entry into economic theory and high transaction costs to acquiring an understanding of it. *Barriers to entry* and *transaction costs* are technical terms from economic theory. There are a lot of them. A barrier to entry is what makes it difficult to get into a market that others are already trading on. When it comes to economics the main barrier is all that math (*maths* to the British reader). Transaction costs are what you have to pay just to make a trade—to buy or sell. Sometimes it's a broker's fee paid to a third party. One of the transaction costs of getting a handle on economic theory is finding your way through the technical terms and equations. One aim of this book is to reduce the transaction costs and remove the barriers to entry. The technical lingo will be explained, but it won't be used much.

There'll be some equations in the chapters to follow. There are two reasons for this. First, I think a lot of people interested in economic theory want to know what the equations say, what they mean. They think that if they don't know the equations, they won't really understand the theory, at least not the way economists do. They are mistaken. But the only way to show this is to walk through the equations. The math in economics is easy to understand. More importantly, seeing how its equations work together reveals a lot about how different economic theory is from other theories that use the same math. So the second, equally important reason for laying out the equations is to show how the math expresses simple ideas and thought experiments that drive the development of economic theory. The equations will all be explained (and sometimes explained away) so that you, the reader, will see what they really add without having to keep them in your head as we proceed. If you really can't stand equations, you can skip them without any real loss.

The reason businesspeople, politicians, and the rest of us think we have to take economics and its math seriously, even when we don't understand it, is because it's *science*. Therefore it can be relied on; in fact it has to be taken seriously. It's the only tool we have to understand and improve the economy.

Most of us have at least a rough-and-ready idea of what a science should look like, and to a first approximation economics seems like one. It certainly looks more like science than the other social sciences do—sociology, anthropology, political science, even a lot of experimental psychology. There's that math, the graphs, and a lot of statistical tools applied to large bodies of data that are constantly updated and woven together into economic models. Plus that Nobel Prize, announced every year along with the ones in physics, chemistry, and "physiology or medicine" (read biology). And then there's the theory. That's what makes economics stand out from the other social sciences. It's what makes economics look so much like a hard science—even more like a hard science than a lot of biology.

The physical sciences—physics and chemistry—are hard sciences not just because all that math makes them hard, as in difficult to master, but also because they share agreed-on fundamental theory. Every textbook about any topic in these subjects expounds the same theory, and all the textbooks present it in roughly the same way, even down to the problems they set at the end of each chapter to test your understanding. The same goes for economics! Economics looks even more like physics and chemistry than most of biology does, just because it has a theory that does the same things for it that the textbook theories do in physical sciences.

But there's a difference between how theory works for physics and chemistry and what it does in economics. A big difference. It's one that matters for how we non-economists should treat economics, and especially how we should use its theory to understand and shape our lives, our societies, and our cultures.

Ever since the scientific revolution of the seventeenth century, physics and chemistry have been data driven. The objective of making sense of the data relentlessly drove theory in these two fields. It was hard, quantitative data that got these sciences going. It has always been more and better data that shaped their theories and reshaped them, sometimes radically, over centuries. The theories in physics and chemistry textbooks all look the same now because experimental scientists pushed the theorists' models into the only shapes that made sense of their data.

None of that is true about economics. Economics is theory driven, not data driven. And it has been driven by the same theory, unchanged in the most significant respects, for two hundred fifty years.

That's the single most important thing you need to know about economics in order to see what it is and isn't good for. Once you understand why the subject is theory driven, a lot about economics becomes clear: why its headline predictions haven't improved over time and why that doesn't matter to economists; why its explanations of the past—distant and recent—are so deeply satisfying to economists despite the persistent weakness of the evidence behind them; and why, despite these problems, economic theory is the most indispensable tool we have for organizing our social lives and crafting human institutions.

What should be troubling about the subject to economists, and what is troubling to the rest of us who consume economics, is just how bad it is at what everyone—including the economists—thinks its job is as a science. This book is not going to belabor the obvious fact that economics is no better at explaining or predicting outcomes than it was fifty, a hundred, or a hundred fifty years ago. Nor will it take on the economists and others who have offered to explain away that track record. If you think that economics is moving forward in the way that a cumulating, improving empirical science should—revising theory in light of data to increase precision and expand the range of explanation and prediction—then you don't need this book. But if you need to know why it is that a theory with such a track record of predictive failures and post hoc explanations for these failures is still with us, keep reading. We'll see why, in spite of what it can't do, economic theory remains in the form it took at its beginning. (*Spoiler*: That's because it gives us the indispensable tool for crafting and fine-tuning the institutions we use to make human society tolerable. Reason enough for us outsiders to get a handle on its theory.)

It's worth saying a little more about how explanation and prediction are related in empirical science. Sciences can't run before they walk or crawl. Physics began to crawl in the time of Kepler, to walk with Newton, and to run with Einstein. Chemistry started to get a few things right a hundred years after Newton, with Priestley and Lavoisier. By the time Mendeleev introduced the periodic table in 1869 it was off and running, and then Bohr handed it most of what it still needed in his quantum model of the atom. Biology was stamp collecting before Darwin. Each of these three subjects began with very limited data domains whose explanations were certified by some successful quantitative predictions. In physics the domain was the planets and a comet; in chemistry it was the behavior of gases;

in biology it was breeding experiments in botany. Available instruments didn't always permit scientists to secure data with sufficient precision and calculate to enough decimal points, to choose between equally good or bad explanations, or to fine-tune the ones that were accepted. In each there was an iterative process of theorizing, then collecting data, and using both to build instruments, which was repeated over and over, resulting in persistent increases in predictive accuracy and improved models, and expanding the domains of data to which an explanatory theory was applied.

The process strongly vindicated the idea that each of these disciplines was on the right track. They were framing successive theories that increasingly approximated the right explanation of the data in their domains, as substantiated by their improving precision and expanding range of successful predictions. Of course, lots of these predictions were of no practical significance in themselves. Often they were just predictions of uninteresting side effects and by-products, now accurate to more decimal places. The process was always data driven, a cycle in which models, theories, even whole paradigms were revised, radically reconfigured, overthrown, and discarded, until after a certain point a single package of them became canonical enough to dominate all the textbooks in a discipline.

The only respect in which economics looks like these subjects is in the hegemony of the theory it teaches across all its textbooks. And the biggest difference between economics and a hard science is that it's still using recognizably the same theory it started with in the eighteenth century, just as physics was hitting its stride, before chemistry really got going, and almost a century before biology found its footing.

In economics there has been no cycle of updating—new data haven't forced revisions in theory that guide new experiments and observations that in turn replace or fine-tune the theory, shaping still more data collection or better instruments to sharpen the measurement of old data. That means economists haven't been able to fine-tune or even test the explanations their theory provides. The explanations economists give "feel right" to them and to those who know the theory. But without any sustained improvement in predictive precision, economics doesn't have the kind of independent evidence of being on the right track that the natural sciences do.

There was a time, almost a hundred years ago, when it looked like economics was going to change its shape in the way sciences do. The profound and long-lasting global depression between the world wars was data! It produced enough dissatisfaction with economics and its ruling theory to result in what historians of science like to call a paradigm shift: the Keynesian revolution's repudiation of "classical" economic theory. The exception proved the rule, however. Subsequent discordance between Keynesian theory and newer data didn't drive economics further forward. It drove economics backward to "new classical" economic theory—the old wine in the old bottles.

Why? That's the updated, twenty-first-century version of the question: Why is a theory with such a track record of predictive failure and explanatory imprecision still with us? The question isn't about why economic theory is hard to get right; economists think they've gotten it right pretty much since Adam Smith in 1776.

The question also isn't why the economy itself is hard to understand. It's easy to catalog the factors that make it multicausal, hard to subject to experiments, expensive to collect data about, and difficult to analyze statistically. These factors should make economics, like all the social sciences, much harder than physics or chemistry or even biology (which is harder than physics). We've made so much more progress in natural sciences than in social and behavioral sciences. We understand nature all the way down to the fermions and bosons, just because the behavior of physical things has been simple enough for us to grasp. Whether we can succeed in providing theories in economics (or the other social and behavioral sciences) as powerful as those of natural science is a good question. Another good question is why we haven't been able to do so yet. But those are not the pressing questions we are going to explore here. Our questions about economics are different: urgent and practical, not historical or philosophical.

Why have we bought into its post hoc explanations through three centuries? Given its track record, should we treat the theory that drives these explanations as a guide to the future? How could it, after all, be indispensable to human social life, now and probably forever?

We can't answer these three questions without learning a certain amount about economic theory. To repeat, that's not the same as learning economic theory—the sort of thing taught in micro- and macroeconomics courses. To learn economic theory, it's best to read one of the textbooks. They're pretty much all the same. What we need is to understand why economic theory looks the way it does, and why it has looked that way, unchanged in substance, since soon after Smith, even as it has acquired more and more mathematical trappings and the *appearance* of a modern, up-to-date, quantitative scientific theory.

Understanding what makes economic theory tick will enable the outsider to see at least three things about economics: why its predictions don't ever seem to get better in the way theory in a quantitative empirical science is supposed to; how much reliance we should place in its after-the-fact explanations of events it didn't manage to predict; and finally, how important, indeed invaluable, it really is when it comes to designing and enhancing the institutions of our civilization. Economics is really important. It's the best and perhaps the *only* reliable thing we have to go on when it comes to organizing social life. And that's true even though economics doesn't work the way a science should, its Nobel Prize notwithstanding. There isn't much reason to believe the details of any of its theories or models. But as a guide to protecting ourselves against the worst ravages of human rapaciousness, it's a godsend. And that's reason enough to take it with the utmost seriousness.

The chapters to follow will walk you through the moves economists make in articulating their theory and the reasons the theory developed as it did. Some of what we'll do is historical, seeing how economists had to reconfigure their theory, mainly to reduce its scope and ambitions, and how their interpretations of it had to change. Once you see that history, you'll appreciate how different economics is from any other discipline in the natural or social sciences. You might even decide it's not an empirical science at all. (Remember, that Nobel Prize in "economic sciences" is mostly a matter of branding. It's officially the Swedish State Bank Prize in Memory of Alfred Nobel, started in 1968, two-thirds of a century after the five original prizes began to be awarded by the Nobel Foundation.)

Some parts of economics are more exciting than others. The cut and thrust of public debate usually focuses on macroeconomics—how changes in interest rates, money supply, and the government budget affect the employment rate and inflation, and how government responds to booms and busts. But in economics nowadays, as we'll eventually see, all the action is really in microeconomics—the theory of individual choice behavior and its consequences. That means that to build our understanding of what's going on in the macro theory that most people are interested in, we have to start with micro theory: how economics treats individual choice, which it calls *rational choice theory*.

Rational choice theory is often ridiculed as the story of *Homo economicus*, or economic man, a relentlessly selfish, lightning-fast calculator who is always looking out for number one. Making fun of this assumption is easy. But economists know as well as their critics what's factually wrong with rational choice theory. The trouble is economics can't improve it, and economists won't give it up. Why that's true is the question we'll answer in chapters 2 and 3. We'll see that the answer goes back to Smith's great insight about the market economy in *The Wealth of Nations*, the work that did for economics what Newton's *Principia* and Darwin's *On the Origin of Species* did for physics and biology, respectively: permanently reorganize the domain. With only slight exaggeration it is said that everything in philosophy is a footnote to Plato. With greater justice it can be said that everything in economics is a footnote to Smith. *The Wealth of Nations* set the marching orders for economics in 1776. It's still calling the tune two hundred fifty years later. That tune begins with *Homo economicus* and ends in contemporary macroeconomics.

The math that economic theory is studded with shouldn't stop people from understanding what's going on. Seeing how and why calculus in particular came into economics and why it sticks around is the subject of chapter 4. The work it does in economics is very different from its role in the hard sciences, as we will see. In this chapter we'll walk through some very impressive contemporary mathematical economics, not exactly deconstructing the equations but showing what they say and how much they contribute to expressing ideas that began with Smith.

Chapter 5 explains why there is no room in the foundations of economic theory for money. Microeconomic theory is like *Hamlet* without the prince. Of course, economists will give you an explanation for its absence. They argue that money is of no real importance in an exchange-based economy among rational agents. Maybe that's right, but it just raises more problems for any attempt to understand macroeconomic theory. The fact is that economists don't even agree on what makes money money. Knowing its history—the trajectory of money's development from the beginning of agriculture ten thousand years ago till now—helps only a little. Money's importance in the real economy is a matter of its future, not its past or even its present. And economics is not good at telling us about the future.

Chapter 6 moves us on to the current state of play in macroeconomic theory and shows how Smith's vision still sets the agenda for the subject. When we're done with the chapter, you'll understand economists' unwavering commitment to just one kind of macroeconomic model, come what may in the real economy. Except for the thirty-year Keynesian interlude half a century ago, that commitment has been so unswerving for so long that contemporary macroeconomics calls itself "new classical" economics. Consumers of macroeconomic theory, especially public and private policymakers, need to understand why macroeconomic theory isn't sensitive to even the biggest economic upheavals, and why its clock has turned back and then stood still since before the subject was dreamed up in the 1930s.

The two most important markets for most of us are the labor market and the financial market. The labor market is important because it's the only market in which almost all of us are sellers, not buyers. We sell our labor in the market in order to be buyers in almost all other markets. So we want that market to work well for us. It doesn't. It can't. And standard economic theory won't help us understand how bad a job that market does for most of us. It lacks the tools to fix it because it can't explain it.

The same goes for the financial market. It's crucial for everyone's future. Its smooth or bumpy operation determines how efficiently the economy will be able to meet all of our insatiable wants and pressing needs in the future. It's even more important for developing economies than it is for developed ones. But the financial market works in ways that are even more of a departure from standard economic theory than the labor market is. They are both domains of what the economist calls *market failure*.

The labor and financial markets are not the only domains of market failure, but they are two of the most important, extreme, and obvious examples. It turns out that all markets fail, to varying degrees, in two respects: they don't behave the way economic theory describes market processes and, more important for economics, they fail to be efficient. They fail to do what economists can mathematically prove markets *should* do best: produce an optimal supply of what people actually want and can afford.

In chapters 7, 8, and 9 we'll see how serious the problem of market failure is for economic theory. We'll do so using the examples of the labor and financial markets because of their importance and the obviousness of their market failures. The problem of market failure and how to explain, predict, and correct it brings us face-to-face with *game theory*, the indispensable tool for doing these three things—explaining, predicting, and correcting real markets.

"Game theory" is the worst name for the most important theoretical innovation in the social sciences since Adam Smith invented economics in 1776. Game theory is not a theory about games like sporting contests, cards, chess, or Monopoly. It's the study of *strategic interaction*—what happens when agents have to make choices, knowing that other agents' choices will affect the outcome of their own. Game theory should have been called the theory of strategic interaction. Whatever we call it, game theory is the indispensable tool for dealing with markets that fail. And almost every market fails, as we'll see, because people are almost always strategic in their choices. So game theory is where (almost) all the action has to be in economic theory.

Once we see this, it becomes apparent why, despite all its problems as empirical science, economic theory is a tool we cannot do without. What we really need economic theory for is *institution design*: figuring out how to craft the practices, norms, rules, and laws that govern human strategic interaction so as to preserve, protect, and defend social welfare and civilized life. This is what economic theory can do for us that nothing else can. Providing us the tools for institution design is so important that it excuses all the problems economics faces when it comes to explanation and prediction.

The problem of institution design goes back to the social contract thinkers of the seventeenth century, especially Thomas Hobbes. He was the philosopher who taught that in the state of nature "the life of man [is] solitary, poor, nasty, brutish and short." The state of nature is the "war of all against all," or anarchy, in which there are no enforced rules to prevent people from doing unto others before others do unto them.[2] The reason Hobbes gave for the awfulness of the state of nature is that in it we have to be selfish, rational egoists, always looking out for number one. Anyone who isn't won't survive.

Hobbes argued there was only one way out of the state of nature: everyone in it has to give up all their power over themselves, and whatever they (temporarily) control, to one and only one absolute sovereign, the Leviathan—the big fish, the king, the lord protector, the dictator. That was Hobbes's solution to the institution design problem of escaping the state of nature. Modern game theory can help us see what Hobbes was getting at.

But no one in the following 450 years has accepted Hobbes's solution to the problem of how to escape the state of nature. The trouble was, their alternative solutions made the problem too easy. John Locke and Jean-Jacques Rousseau just assumed that in the state of nature we aren't selfish egoists. We can afford to be and in fact are nice to each other at least often enough that we can agree on a limited government to control us, instead of an absolute dictator.

Locke and Rousseau were too optimistic about human nature. Time and tide have taught us that it's wisest to assume the worst-case scenario. Better to suppose that people really are self-interested and selfish preference maximizers, just as economics presupposes. Of course, we aren't all rational egoists, *Homo economicus*. That's why economics gets things so wrong. But enough of us are, enough of the time, that it's more than prudent to try to protect ourselves from ourselves.

We need to construct institutions—laws, rules, norms—that will insulate us from the worst depredations of the egoists among us. Maybe we can even design institutions that will harness everyone's limited self-interest to promote everyone's mutual advantage even when we aren't all aiming at it. Economists have a name for such institutions: they are *incentive compatible*. An institution is incentive compatible if it includes carrots and sticks that harness people's self-interest to make them act in accordance with the institution's objectives. Such *mechanisms*, as economists call them, encourage, nudge, or if necessary force people to do the right thing by making it in their interest to do so. Speed cameras and good-driver insurance discounts encourage people to obey traffic laws. It may sound obvious to us that institutions shouldn't create perversive incentives. But the game theorists who turned this insight into a tool of institution design have begun winning a lot of those Nobel Prizes in economics.

It's very hard to design mechanisms that don't give at least some people perverse incentives. It's hard to design foolproof incentive-compatible institutions because some people are always tempted to find loopholes, secure exceptions, or risk being caught trying to game the system or work around the rules. Hobbes called these people "fooles," and Hume described them as "knaves."[3] Both were at a loss about how society can protect itself against them. But Hobbes and Hume lacked game theory. That theory is the only guide we have in the constant struggle to design and redesign institutions to protect ourselves from the selfish egoists in our midst. There may not be many rational egoists. There are, thankfully, even fewer sociopaths. But there are always enough of both to make the job of defending against them unending—an arms race, in game theory lingo. Helping us win arms races, or at least not lose them too badly, is something that economics can do because of its unswerving commitment to always thinking about what *Homo economicus*—the selfish, rational egoist—will do.

2 Economics Won't Budge, Can't Budge, from Rational Choice Theory

Economic theory starts with rational choice theory, or RCT for short.[1] From the simplest case of the law of supply and demand all the way to the furthest reaches of macroeconomics, rational choice theory is the engine that drives economic theory. In one form or another, it has been at work for 250 years, from the beginning of economics in Adam Smith's *Wealth of Nations*. All of the insight and illumination the theory confers starts from RCT.

But rational choice theory doesn't do its theoretical work in the way we'd expect of a mathematical model sitting at the foundations of an empirical science. In this chapter we'll see something troubling about RCT as a scientific model—something that goes on to haunt the rest of economics, since RCT is the beating heart of its theory. In chapter 3 we'll see what RCT does for economic theory and why it will never be given up as the theory's starting point, in spite of its troubling character as a description of human choice.

Smith appreciated that individuals are self-interested, seeking to maintain and increase their well-being, satisfaction, or happiness. This, he recognized, is the source of people's "propensity to truck, barter, and exchange."[2] *Truck* as Smith used it was a synonym for *barter*, one preserved in the later English expression *truck-farm*. He famously conjectured that self-interest, operating through the willingness to exchange, results in benefit to the whole society through the operation of the invisible hand of the market:

> Every individual necessarily labours to render the annual revenue of the society as great as he can. He generally, indeed, neither intends to promote the public interest, nor knows how much he is promoting it.... *He intends only his own gain*; and he is in this, as in many other cases, led by an invisible hand to promote an end which was no part of his intention.... By *pursuing his own interest*, he frequently promotes that of the society more effectually than when he really intends to promote it.[3]

For the next 150 years or so, economic theory sought to convert Smith's conjecture into proof that markets attain what Smith called society's interest. RCT was the tool it needed to do so.

RCT: The Pure Theory and the Messy Reality

The effort to reformulate Smith's conjecture to give it a knockdown argument begins by formalizing self-interest into a theory about rational choice. Remember, *rational choice* is a technical term that bears some relation to the ordinary meaning of the words, but it has been designed to deliver some important insight into the working of the economy. Memorizing the technical meaning won't be required, but taking rational choice apart and seeing what makes it tick will make all the difference throughout this book.

First economics defines rationality; then it postulates that individual economic agents—buyers and sellers, firms and households, producers and consumers—are rational, that they behave according to the definition. There are four conditions that agents have to satisfy for their choices to be rational:

1. Agents rank all available alternatives as more preferred, less preferred, or tied for preferability (what economists call *indifference*).
2. Agents are insatiable; they prefer more of anything they want to less of it.
3. Agents' preferences are *transitive*: if they prefer A to B and B to C, then they prefer A to C.
4. Agents always choose the most preferred among all available alternatives.

Economics defines rationality as behaving in ways that satisfy these four conditions. That's all there is to RCT: these four axioms, assumptions, postulates. Why these four? Each of them is needed to prove the most important theorem in mathematical economics—the formal proof of Smith's conjecture that the invisible hand of the competitive market harnesses every individual's rational self-interest to produce the best overall outcome for society. To get as close as possible to converting Smith's conjecture into a mathematical proof requires choosers satisfy these four conditions. We'll see how it does so in chapter 3.

But economics needing to define rationality this way to get some mathematical result doesn't seem to be enough of a reason to hypothesize that people really behave this way—not if we're interested in explaining and predicting people's economic actions. That's the source of the criticism of economists' unreasonable attachment to *Homo economicus*.

The trouble with these four claims, as you must have noticed, is that they are all false. It's easy to find exceptions to all of them in your own experiences.

The first condition is far beyond the ability of any human to satisfy. It implicitly requires two kinds of omniscience. First rational agents have to know all the available alternatives (and the probabilities of each one occurring), and second the agents must know their own preference rankings among them all. It's obvious we don't know all the alternative courses of action open to us. Maybe it's slightly less obvious that we don't know how we feel about all of them. Can you really rank every alternative you can contemplate as more preferred to, less preferred to, or exactly equally as preferable as every other one? Which do you prefer, cucumber ice cream or coffee-flavored salad dressing? How do you rank being drawn and quartered compared to being burned at the stake? Do you really like one better than the other? Or are you exactly indifferent between them in your preferences, so it doesn't matter which method of execution you face?

The second requirement, insatiability, is pretty weird. Are you insatiable when it comes to tequila shots? Isn't there a point after which you don't just want no more, you want fewer than you had?

Number three, transitivity of preferences, seems reasonable as a description. But testing it in real time, which is the only way we can see if it holds, requires memory of alternatives and no changes in taste. It's easy to get people to violate transitivity simply by describing alternatives in different words. And then there are the transitivity violations people commit because they aren't good with numbers, odds, probabilities, and bets.

Since we almost never know what all the available alternatives are, we all sometimes violate the fourth condition. Even when we know all the likely available alternatives, we sometimes make mistakes: we reach for the wrong available alternative even when we've thought about it.

These four conditions on rationality are very stringent. By the standard of RCT, people aren't always rational. At least sometimes they are irrational, and only rarely are they fully rational. As a description of actual individual human choice and behavior, RCT is pretty poor.

It Doesn't Matter; RCT Is a Model!

People certainly aren't always perfectly rational by economists' definition. So what? RCT is a *model* of choice. Like other models in science, RCT involves idealizations—harmless ones that need to be made if we are going to use it to explain and predict economic behavior. That has been economists' response to criticisms of *Homo economicus* for almost a hundred years.

The trouble with this take on RCT is that economics doesn't really treat it as an explanatory or predictive model, not the way other sciences use them. Its role in economic theory is much more fundamental than the role of idealized models in empirical science.

To see the difference between RCT and idealized models in science, let's go back to one we all learned in high school: the kinetic theory, or the billiard ball model of a gas. The next couple of paragraphs may seem to be a digression, a trip down memory lane, but we'll be coming back to this idea of a scientific model over and over in the rest of this book. I promise the digression will pay off.

High school chemistry begins with data collected as far back as the 1600s by British and French chemists about the temperature, pressure, and volume of gases. Boyle's law tells us that the volume of a gas is inversely related to its pressure. His data are shown in figure 2.1.

Charles's law was the next gas law to be discovered. It states that the volume of a gas is directly related to its temperature. Then Gay-Lussac discovered

P (in Hg)	V (in.^{3})	PV
12.0	117.5	1410
16.0	87.2	1400
20.0	70.7	1410
24.0	58.8	1410
32.0	44.2	1410
40.0	35.3	1410
48.0	29.1	1400

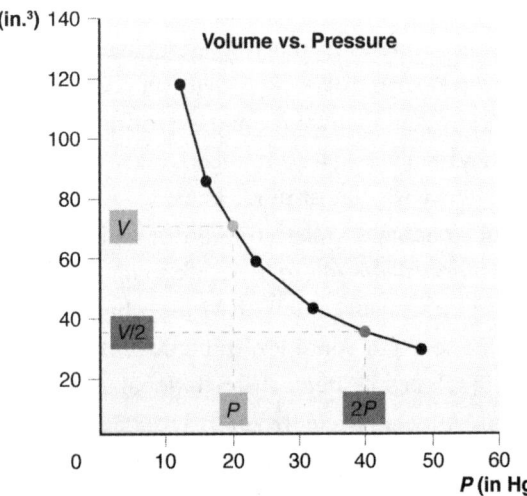

Figure 2.1
The graphs plot real data points in real units from experiments that have been replicated for three hundred years.

that when holding volume constant, pressure is directly related to temperature. Of course, chemists were able to measure temperature, pressure, and volume with considerable accuracy. They used the relationships they discovered to predict with equal accuracy the amount of change in each variable that resulted from a change in one of the other two. Throughout the nineteenth century data were driving the formulation of several more gas laws. They were eventually combined into the ideal gas law: $PV = kT$, where k is a constant. Why *ideal*? We'll get to that.

One of the most profound achievements of nineteenth-century physical science was the formulation of an ideal model that explained the gas laws. This was the kinetic theory or billiard ball model of a gas. This finally revealed what heat really is: molecular motion. Assume that gases are composed of point particles, with mass but no volume, and that they bounce off one another and the walls of a container frictionlessly, losing no momentum. Add the assumptions that the temperature is a measure of the average kinetic energy of the gas molecules (i.e., one-half their mass multiplied by their average velocity squared: $\frac{1}{2} mv^2$) and that the pressure depends on the momentum of the molecules bouncing off the container walls. With these two assumptions, you can mathematically derive all of the gas laws by applying Newton's laws of motion to the molecules. So we explain the

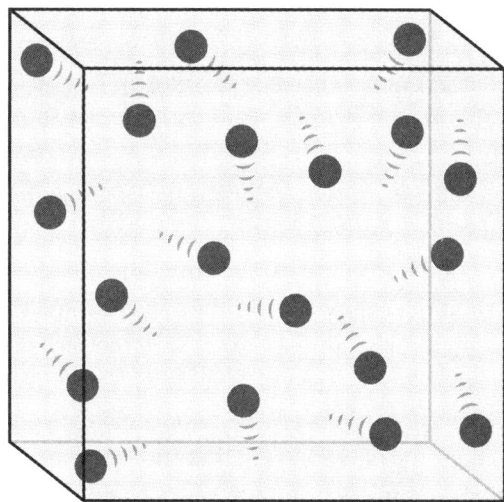

Figure 2.2
The billiard ball model of gases.

macroscopic properties of a gas—its temperature, pressure, and volume—by deriving them from the properties of their microscopic constituents, the gas molecules. The development of the billiard ball model of gases was the biggest step taken in the nineteenth century toward revealing the existence of atoms and molecules. If there had been a Nobel Prize back then, Lord Kelvin would have won it for using this model to show that heat is just the motion of molecules.

The billiard ball model makes several idealizations so extreme that we know from the get-go that its assumptions are false. First it assumes that molecules are point masses: particles that have mass but take up zero space! Impossible, according to Newtonian physics. Equally impossible, these molecules are assumed to bounce off each other and the container wall frictionlessly. And the model assumes the molecules don't rotate or vibrate. These contrary-to-fact assumptions are what make the billiard ball model an idealization. The ideal gas law is so called because it is a law about gases composed of ideal molecules—point masses that move frictionlessly. But we can derive $PV = kT$ mathematically from the billiard ball model.

Here's what makes idealizing models so important to science. When we complicate the model by reducing the idealizations, making the model more realistic, the model begins to show us how to revise $PV = kT$ to make more accurate predictions. Making the model more realistic—less ideal—broadens the range of its predictions to more and more values of pressure (P), volume (V), and temperature (T). Adding variables that the original model idealized away enabled chemists to apply the model to predict P, V, and T for gases that hadn't even been known about when the ideal gas law was formulated.

The ideal gas law has many exceptions, especially at extremes of pressure and volume. It doesn't work well for gases that are very light (hydrogen, for example) or very heavy (like radon). But thinking about the idealizations in the billiard ball model helps explain these exceptions. There's a limit to how closely gas molecules can get pushed together as volume decreases, especially heavy or ionically charged gas molecules. That will cause a breakdown in the neat relationship between volume and temperature.

In thinking about the data showing that the departures in the behavior of real gases could be explained and predicted, chemists kept changing the assumptions of the billiard ball model. That's how the model was improved over the nineteenth century and into the twentieth. Making more realistic

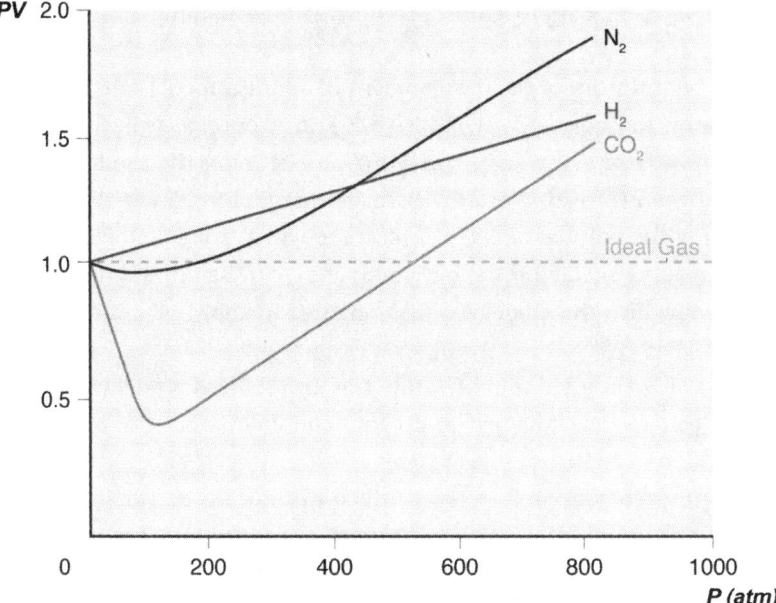

Figure 2.3
Real gases depart from the ideal gas law.

assumptions about the size and intermolecular attraction or repulsion of molecules complicated the simple formula. Together with more data, these complications turned the billiard ball model into something less idealized and more powerful as an explanatory and predictive tool of chemistry. By the end of the nineteenth century the simple equation had given way to something more realistic and much more predictively accurate: Van der Waals's equation, $(P + a/V^2)(V - b) = kT$, where a measures the attractive forces between molecules, b measures the repulsive forces between them, and V is the volume of the container. If a and b are zero, the Van der Waals equation reduces to the ideal gas law.

Notice that if you have very good data on the actual values of P, V, and T for a particular gas, you can use the equation to work back to values for a and b, the unobservable properties of the gas molecules. Eventually, advances in physics enabled scientists to measure a and b more directly, thus independently confirming the model's predictions and also establishing the reality of atoms. The models and data work together: data refine the model; the model predicts new data. Repeat for a hundred years and you get something approaching a complete understanding of the behavior of gases.

This brief history of the development of an ideal model into more realistic ones is the story of how models work in science: a cyclical process of model predicting, new data, correcting and refining the model, predicting, and then correcting again. There are many examples of this cycle of improving the number of decimal places to which a prediction works out, and of making predictions about completely new samples different from the one to which the model was first designed to apply.

It would be great if we could treat RCT as an idealized model just like the billiard ball model, designed to do for economics what the billiard ball model did for chemistry. So let's try.

RCT's most obvious role in economics is to allow the economist to derive downward-sloping demand curves from the preferences of individual buyers on a market. Sellers are also assumed to be rational and RCT can be used to build the upward-sloping seller's supply curve. We'll see exactly how you get downward-sloping demand curves from RCT in the next chapter. If RCT works like the billiard ball model, then by deriving demand curves from RCT we explain them the way deriving $PV = kT$ from the billiard ball model explains the gas laws.

But we have empirical data for the gas laws. The data are precise, stable, and replicable, and as instruments got better and better, we were able to engage in the iterative process of data collection, model refinement, more data collection, more refinement . . . This is where the difference between RCT as an ideal model and the billiard ball model emerges.

The actual datasets of how demand for any commodity varies with its price don't show the nice precision and stability that the gas laws manifest. And even if they did, you couldn't derive any of that detail from RCT. Sure, the curves of data points for actual demand are almost always downward sloping. But their slopes change as their prices (and the prices of many other things) vary, and they do so from day to day, indeed from moment to moment. (Just look at Uber's hour-by-hour data about how demand drives its "surge" pricing.) The most we can get out of RCT is that demand curves slope downward. Exactly how steeply each of them slopes and for how long they stay still is something RCT can't deliver. If RCT could give us some numbers, even wrong ones, then we'd be able to get into the business of refining RCT, adding variables, reducing idealization, and complicating the model but getting a better grip on the data. This would increase our confidence that we were on the right explanatory track, in spite of the idealization RCT imposes on choice.

Behavioral Economics to the Rescue?

But wait, isn't the impressive and illuminating discipline of behavioral economics doing just what RCT needs—identifying where people "go wrong," make mistakes, and violate the dictates of RCT? By incorporating what it has discovered about the foibles of real human choices into RCT, can't behavioral economics improve RCT like chemists improved the billiard ball model over time?

Behavioral economics staked out exactly this research program in the laboratory by setting up controlled experiments with people to see when and how their choices violated RCT and then improving the RCT model by adding components that would make better predictions beyond what the experiments showed. This is the iterated process that is familiar from experimental science.

Did it work? Well, it produced a lot of interesting findings, and it won a couple of Nobel Prizes for social psychologists and an economist. But nothing that behavioral economists discovered about choice had any impact on economists' reformulations of RCT or its role in economics. Let's see why.

One way to think about how behavioral economists proceeded is to focus first on how well RCT works in a casino, where all the probabilities are known—all the odds and payoffs are posted. In the casino RCT is the way to go. Obeying its axioms strictly will, on average, win you the most money, or at least enable you to play longer before you lose it all. (In a casino the odds are always stacked in the house's favor.) This is because RCT agents will apportion their bets to the probabilities.

The RCT agent will have no preference (they will be indifferent, in RCT lingo) between a 100 percent chance of winning $1,000 and a 50 percent chance of winning $2,000. That's because the monetary value of a bet is the winner's payoff times the probability of winning. The RCT agent will choose a 50 percent chance of winning $2,001 over a 100 percent chance of winning $1,000 every time. (Notice that you probably wouldn't make the same choice.) Faced with a certain loss of $1,000 or a 50 percent chance of losing $2,000, the RCT agent is indifferent. (Again, this isn't how you'd feel, is it?) Increase the certain loss to $1,001 and the RCT agent prefers the 50 percent chance of losing $2,000. (You don't agree? Are you irrational? You are RCT-irrational.)

RCT has what economists call *normative force* for self-interested agents: those who prefer more money to less money. It tells you what you *should* do to be a rational gambler. In the long run, you'll make the most money and lose the least money if you obey RCT. But one of the first things cognitive scientists discovered while testing RCT on real people (sometimes with real money and sometimes by asking how they'd choose) is that most people they tested were risk averse and loss averse in a way that violates RCT. Almost certainly you are, too.

This means that to explain actual choice the RCT model had to be modified. Daniel Kahneman (and his late collaborator Amos Tversky) proposed to add to RCT a value function that would change the computed value of a gamble (payoff times probability) by some number that increased more steeply for losses than for gains and so reflects people's aversion to loss and risk.

Even before the discovery and incorporation of loss and risk aversion into RCT models, it had been discovered (by animal behavior psychologist Richard Herrnstein) that lots of birds and mammals, including almost

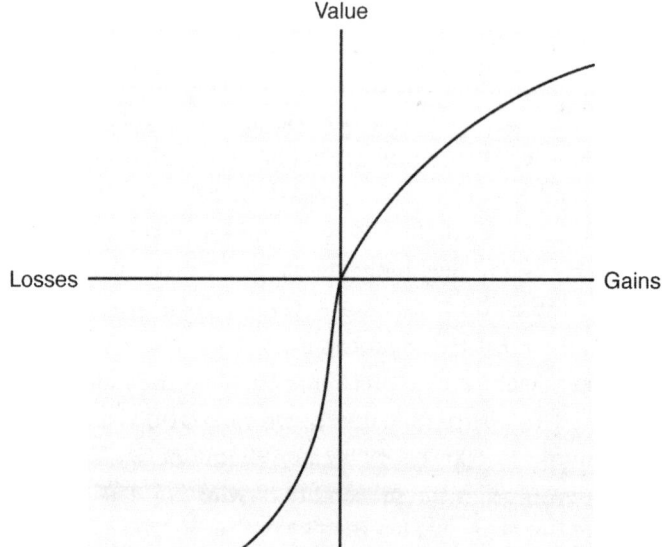

Figure 2.4
The value function leads people to prefer avoiding loss over risking a chance to gain more.

all humans, have preferences that are time inconsistent, violating RCT to greater or lesser degrees depending on how far into the future the payoff is delayed. Given a choice between $10 today and $20 a week from now, many people will take the $10 now. Give them the same choice postponed a year—$10 in a year versus $20 in a year and a week—and people prefer to wait the extra week for the $20. People discount the cost of waiting much more for consumption right away than they do for consumption delayed for longer. By contrast, the RCT agent discounts the future *evenly*—any one-week delay of a future payoff will result in the same choice. If we were to draw a curve of how strong the discount for waiting is, it would drop steeply, like a hyperbola, in the near future and then flatten out for longer delays. So this violation of RCT has come to be called *hyperbolic discounting*. To modify RCT to produce hyperbolic discounting, we add another coefficient that makes today's preference for postponed consumption drop fast at first and then much more slowly as postponement time increases.

Still another famous bit of behavioral economics emerged from experiments in a large number of Western and non-Western cultures. Experimenters

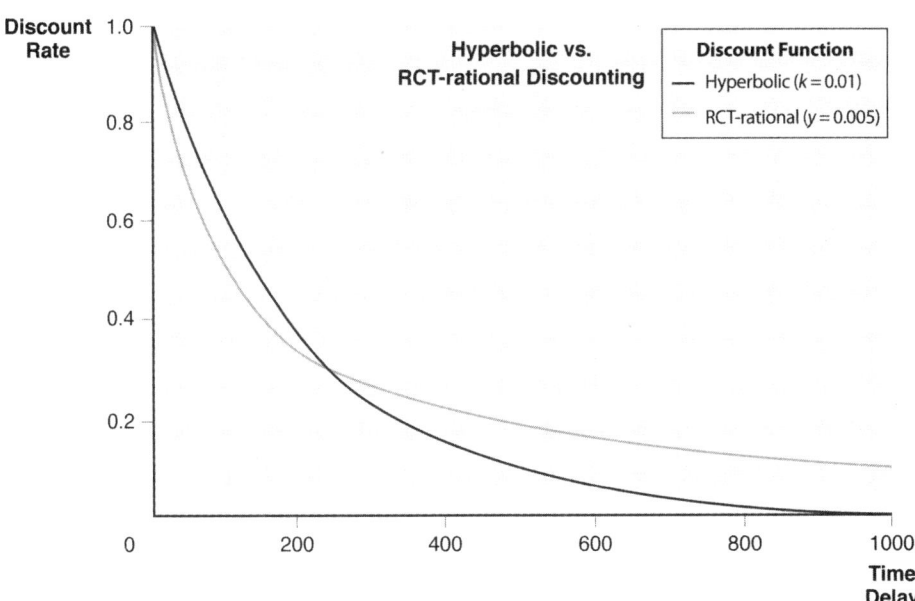

Figure 2.5
Hyperbolic discounting.

set up games in which people have opportunities to do what rational egoism dictates or instead to act in the interests of others. Participants in these experiments were provided with real money, sometimes substantial sums. Then they were asked to play their stakes in games in which they could choose selfish or unselfish strategies. For example, player 1 must choose how much of $10 to give to player 2 in dollar increments. If player 2 refuses to accept the split that player 1 proposes, neither gets anything. RCT tells player 1 to offer $1 and player 2 to accept it. But this hardly ever happens. Almost everywhere in the world this game has been played by experimental subjects, the average offer by player 1 has been $4 or more, and lower offers usually aren't accepted. Player 2 would rather get nothing than take the lowball offer.

Debriefing players after the games suggests there is a fairness norm at work in their calculations. One way of modeling this behavior is by a further complication of RCT that adds preferences for norm-following, or being kind to others. Another is to add preferences for punishing unfair players, even when punishing is costly to the punisher and the punisher isn't the one who has been harmed.

These are only the most famous three of literally hundreds of findings about how people violate RCT because of cognitive biases.

Adding these details improved RCT as a psychological description of the process that produces choices. Unfortunately, the increased realism didn't bring significant improvement in RCT's predictive power. All it has done is provide more retrospective explanations for the myriad circumstances in which RCT fails to predict behavior accurately.

Loss and risk aversion make intelligible the behavior of people who buy "too much" insurance. But it can't predict who will and who won't, or how much extra they'll buy. That's because it's no easier to identify the exact or even approximate shape of Kahneman's value function than it is to identify exactly how it operates on the preferences. We can be pretty certain that there is a wide range of loss- and risk-aversion value functions operating among the range of real economic agents. But we can't measure these value functions, figure out the exact shape of the curve for one person, or average them out for enough people, so we can't tell how much the explanations that add the value function to the original model improve its predictions to show whether we're on the right track. And even if we could do any of these things, economic theorists would have no interest in such empirical findings. Why not?

It's not that behavioral economics has been of no interest to economists. Behavioral finance has made use of hyperbolic discounting to explain well-known phenomena like people failing to save enough for retirement and even addictions like smoking. But calculating the formula for the slope of the hyperbola is no easier than finding the value function for risk and loss aversion. If we could ever find it for one person and one set of choices, it would probably be different for the same person and different choices, not to mention different people and different choices. Figuring out how to fit it into the RCT model is still another problem.

Most behavioral economists accord RCT *normative force*. They accept that the "right" choice, the prudent one, is the choice RCT dictates. But people are irrational. They leave money on the table, as economists see it; if they did the right thing (according to RCT), they'd make more or lose less. Proceeding this way, behavioral economists hope to "correct" people's choices, to nudge them in the direction RCT demands. This might improve the RCT model's predictive power by changing people's behavior instead of changing RCT. The cognitive biases behavioral economists have uncovered show us how to compensate for people's failure to adhere to RCT, not its failure to describe their behavior.

There are other models of choice that have been advanced in behavioral economics that don't start by treating RCT as the baseline right way to choose. Experiments and surveys suggest to some behavioral economists that people making choices don't consciously employ anything like RCT at all. So, they argue, the model of how people choose shouldn't start from RCT and add qualifications to it.

These scientists treat RCT as a failed model of the psychological process of choice and set it aside altogether. They seek a different causal model for the process of choosing and use it to make predictions they hope will outperform those of RCT. They are inspired by the work of Herbert Simon, the first of the behavioral economists to win the Nobel Prize in economics. Simon offered *satisficing* as an alternative to the optimizing that RCT attributes to choosers. Satisficing is a simple heuristic device, a shortcut, a rule of thumb: choose the first available alternative that is above some minimal level of satisfactoriness. At least sometimes choosers are consciously aware of satisficing as the causal mechanism of choice (if conscious awareness is reliable). Following Simon, some behavioral economists have sought other "fast and frugal" heuristics, rules that people actually use in making choices in real time when deadlines loom and information is limited. These scientists hold their models to the standard of doing at least as well as RCT in predicting behavior, and in maximizing gambling payoffs, too.

It's fair to say that the way behavioral economics has developed shows it isn't really a compartment of economic theory at all. It's a project of cognitive scientists. What these behavioral scientists sought and still seek is a model of the mechanism that governs human choice in the brain: what psychological processes, comprising what kinds of computations, give rise to individuals' choices.

Taking RCT as a starting point was a natural step for cognitive scientists, mainly because there wasn't any other model to start with. What psychologists have discovered about human choice may have surprised economists. It certainly has been applied to explain some patterns of behavior we already knew violated the prescriptions of RCT when people make financial and other gambling decisions.

But those who added bells and whistles to RCT to try to bring it more closely in line with actual behavior haven't done what economic theory needs. They haven't been able to make its models useful for improving economics' quantitative, predictive understanding of demand curves or of market processes resulting from people's choices taken more than one at a time. No surprise there. Each person is going to have a different loss- and risk-aversion curve, a different discount hyperbola, and different amounts of weight attached to norms of fairness. These values may change for each person over many different timescales. We'd need ways of measuring these values, and then ways of summing or averaging them over populations, revealing how they change with variables like age, gender, or culture. Even if behavioral economics could do these things, there is no obvious way these factors could be packed into the RCT model.

What all this suggests is that behavioral economics is not going to do for RCT what empirical research in chemistry and physics did for the kinetic theory of gases. There, improving realism and reducing idealization established the explanatory power of the billiard ball model of a gas. It did that by identifying causal mechanisms more fully, which in turn improved the predictive accuracy of successive versions of the ideal gas law. This pattern of turning idealized models into more realistic explanations and more powerful predictive tools is a common feature of empirical science. It's what vindicated the practice of starting out with models we know are false: the billiard ball model of the gas, Bohr's solar system model of the atom, the Punnett squares of Mendelian genetics, the Hodgkin–Huxley model of the neuron as an electric circuit. Each was an idealization that could be improved upon.

Behavioral economics isn't going to underwrite economists' embrace of RCT as this kind of scientifically fruitful model, one that initiates a long-term cycle of improvements vindicating the original idea as the first step along the right track. So behavioral economics makes the problem of understanding economists' attachment to RCT harder rather than easier. It shows that we can't treat the RCT model as a garden-variety idealization in a scientific research program. RCT isn't the first step along a path toward uncovering causal mechanisms and thereby enhancing the prediction of data and the design of instruments that further advance economic theory's research program.

Of course, behavioral economics is not itself undermined as a significant research program by this failing. One or more of the many models that behavioral economists have advanced may well be vindicated in just the way further research has sustained the relevance of highly idealized models in other sciences. Let's hope so. We need to understand the psychology of individual choice—even if economic theorists don't think they need to.

Economists Insist We're Missing the Point

Most economists will be pulling their hair out by now, loudly shouting that we've missed the point of RCT completely. They will insist that we have been engaging in a wild goose chase entirely irrelevant to economics and to the role of RCT in its theory.

The economist's RCT isn't a psychological model at all! they cry. Hasn't been for a hundred years! To repudiate RCT as a psychological theory, economic theory helped itself to a couple of intellectual fashions that reigned in the social sciences for much of the twentieth century. As intellectual fashions changed, the interpretation of RCT also changed to keep it safe to use in its original form as an idealized model, even if it was one that couldn't be improved.

To begin with, starting in the nineteenth century, economic theory rejected introspective psychology as a source of evidence against RCT or a source of data that might help improve it. Appeals to introspection as insight into choice were labeled fallacious "psychologism." Once behaviorism came into full swing among psychologists in the 1920s, it was even easier for economists to insulate their discipline from contact with cognitive processes. Following the future Nobel Prize winner Paul Samuelson, by the 1940s economists had adopted *revealed preference* as their official interpretation of RCT. "Revealed preference" is a bit of a misnomer. It doesn't mean revealing the psychological preferences that drive consumers' actual choices. Instead it labels a construct that economists can build up out of people's actual choices—their observed behavior. If someone's actual behavior when choosing among a number of available alternatives obeys the four requirements for RCT rationality, then the economist can build up a *preference ranking* for the chooser. This ranking is the chooser's revealed preference ranking—a summary of how the chooser would select among actual and possible bundles of commodities. Revealed preference rationality is, of course, just as idealized as RCT when treated as a model of the psychology of choice. But at least it was observationally testable and couldn't be held hostage to findings and theories about human psychology.

Even after behaviorism began to lose its luster for psychology, it continued to charm economists due to the way it insulated RCT from experiments with human subjects designed to identify the psychological causes of choice behavior. But if RCT doesn't model choice, what does it do for economics? Conveniently for economics, an answer to this question seemed to blow in on the wings of an intellectual trend that succeeded behaviorism.

Logical positivism treated scientific models and theories as useful instruments for organizing data. It refused to take unobservable objects (like mental states) seriously as making a contribution to the predictive power of models and theories. So it rewrote such models and theories as convenient summaries of what could be observed. It treated explanation as nothing more than the flip side of prediction, without according to it anything that even hinted at providing understanding. In 1953 Milton Friedman channeled this philosophy of science into his essay "The Methodology of Positive Economics," which became the sword and shield of every economist's vindication of the discipline for the next fifty years. Never mind if economics didn't work anything like Friedman's essay confidently asserted it did.

Friedman famously argued that the degree of idealization of a model was irrelevant to its usefulness, and its usefulness was purely a matter of the accuracy of the predictions it could make: "Viewed as a body of substantive hypotheses, theory is to be judged by its predictive power for the class of phenomena which it is intended to 'explain.'" The class of phenomena economics intended to "explain" via RCT was the behavior of markets, not the calculations that drive individual economic choice—that's psychology, not economics.

Friedman put scare quotes around "explain" because he held that the only thing that mattered in science was prediction: "A hypothesis is important if it 'explains' much by little, that is, if it abstracts the common and crucial elements from the mass of complex and detailed circumstances surrounding the phenomena to be explained and permits valid predictions on the basis of them alone."[4]

As a take on the role of idealized models in empirical science, this view is at worst guilty of pardonable exaggeration. If RCT worked like the idealized billiard ball model of a gas, we might sweep aside the objections of a philosopher of science to Friedman's crude overgeneralization about models and theories. But the trouble is we already know that RCT won't pass this test of improving predictive power over iterations that reduce its idealization. Economic theory is no better at predicting human behavior now than it was when Friedman wrote his famous screed in 1953.

Even a sophisticated version of empiricist philosophy of science won't excuse the idealization of RCT if the model isn't quantitatively confirmed by data drawn from observation or experiment. In the end, Friedman's famous polemical paper left unanswered the question of how an uncorrected, unimproved, unrealistic, unpredictive model can be an indispensable tool of economics.

Logical positivism, like behaviorism before it, came and went as a fashion among economists, but the question of what RCT does for economic theory remained. Soon after Friedman's essay, however, his student (and subsequent Nobel Prize winner) Gary Becker produced a clever economic argument to show that RCT has a useful role in economics regardless of its truth, idealization, or realism. In "Irrational Behavior and Economic Theory" he tried to show that whether or not RCT was right in its assumptions was irrelevant to its role in economic theory.

Becker began by announcing the real interest of economics, what he called the *rationality of markets*. He then diagnosed the distraction that resulted from taking the RCT model seriously as a description of choice behavior: "Although economists have typically been interested in the reactions of large markets to changes in different variables, economic theory has been developed for the individual firm and household with market responses obtained simply by blowing up, so to speak, the response of a typical unit. Confusion resulted because comment and analysis were directed away from the market and toward the individual, or *away from the economist's main interests.*"[5] But what exactly is market rationality? It's downward-sloping demand curves and upward-sloping supply curves. This is what Friedman held to be the class of phenomena that the model of rational choice and the theory of the firm are supposed to predict.[6] If Becker could show that economic agents, whether rational or irrational, are forced by the market to respond in ways that look rational by the light of RCT, then we could help ourselves to the RCT model as a shorthand device for summarizing how markets optimize behavior, whether or not individual people do.

Think of three types of consumers. One type is rational; they purchase the quantity of a good that maximizes preference. The second chooses how much of a good to purchase by a completely random process—throwing dice. The third chooses, among the available quantities of a good, the amount closest to the quantity chosen the last time a choice had to be made. The second and third consumer types are irrational. They are, in Becker's words, *impulse* consumers and *inertia* consumers. Standard economic theory tells us that when the price of the good in question rises, the rational agent buys less of it, and that's why demand curves slope down—the lower the price, the more is demanded; the higher the price, the less is demanded.

But when the price of a good rises, each of the irrational agents can afford less of it, given their fixed budgets. The impulse consumer chooses amounts by flipping coins or throwing dice. The inertia consumer buys as close to the amount purchased before at the lower price as they can. Given the way they choose the amount to purchase, both will have to purchase less at the higher price. A simple graph shows why.

In figure 2.6, the x- and y-axes represent the amounts of two goods a consumer can buy. The two straight lines, AB and CD, are *budget lines*. They give the maximum combination of each good, x and y, that the consumer can buy at a given price. The steeper line, CD, means that x is more costly in terms of y than it is on the flatter line AB. Assume the rational agent consumes at point a, the midpoint of the budget line AB. Since a is the midpoint of the budget line, impulse consumers will purchase amounts on the AB line to the left and right of a in presumably a normal bell-shaped

Economics Won't Budge, Can't Budge, from Rational Choice Theory

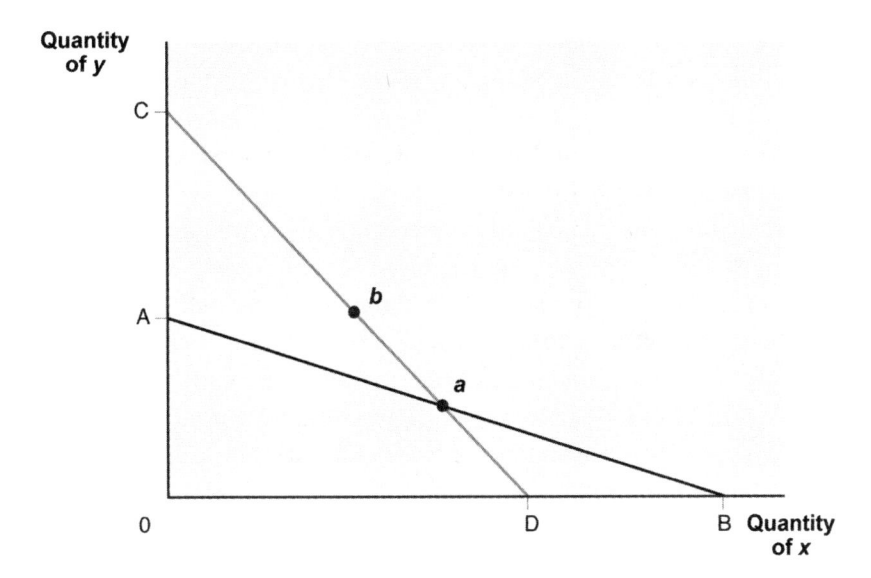

Figure 2.6

distribution curve. Inertia consumers will be located as close as possible to where they were the last time they made purchases of x and y, somewhere, anywhere, inside the triangle A0B.

Now, raise the price of x in terms of y. That is, shift the budget line to CD, a steeper line, which indicates that x is now more expensive relative to y than it was when AB was the budget line. Rational agents have to move their consumption of x and y to the left along CD, since x is now more expensive. The rational agent's demand for x has fallen. Let's put it at b, in the middle of the new CD budget line.

If we look at the graph, we'll see that the impulse consumers and the inertial consumers must do the same thing, even as they choose nonrationally.

The sizes of the two triangles in the graph make this clear. The smaller one is no longer available for either type of irrational consumer to make their choices in. No one is offering amounts of x at the prices inside the smaller triangle. So they can't land there. Impulse and inertial purchases will have to be in the larger triangle in the graph, and above the AB budget line, indicating smaller purchases of x. The inertial consumers who were in the smallest triangle before will have to move to somewhere in the largest triangle (C0D) as close to where they were in the small triangle as possible. They can't be in the small triangle anymore. Impulse consumers will now be distributed randomly around point *b* on budget line CD. Both types of consumers will be purchasing less of *x* at the higher prices, just as the law of demand requires: when the price of *x* rises, the amount the market purchases of *x* declines. It doesn't matter whether the change is the result of rational choice or consumers' impulsivity or inertia.

Bottom line: There's no need to defend RCT as an ideal model of actual human choice. Economic agents can be any mixture of impulsive, inertial, or rational choosers. The laws of supply and demand will still hold. So economics doesn't have to worry about whether RCT is even an approximately correct description of economic choice. That's not what it does for economics. It's just a handy way of expressing the main business of economic theory—that demand slopes downward and supply slopes upward—otherwise known as market rationality.

But then, why give RCT some special foundational role in deriving the laws of market rationality? Why not just start with those market laws and do economic theory without RCT or any hypothesis about how individuals choose? Good question. The answer begins to become clear in the next chapter.

It's safe to say that economic theory did not accept Becker's invitation to treat RCT as just a device for keeping in mind that demand curves decline and supply curves rise. In fact, neither did Becker. He spent the rest of his distinguished career treating RCT as the most important explanatory model in all of social science. Indeed, he proselytized for what came to be called *economic imperialism*, the extension of RCT models of human choice to explain a vast range of phenomena in many other social sciences, especially sociology, anthropology, and political science. He spent the years after writing this paper using RCT to "explain" as much about human behavior, including choices outside of economics, as he could. Fifteen years after he wrote his irrationality paper, he published a book, *The Economic Approach to Human Behavior*, in which he tried to show that RCT explains lots of nonmarket choices, including racial prejudice; the decision to marry, divorce, or have children (and how many); crime; travel; migration; health; and any area of human life in which the individual can weigh ends and means.

Becker even developed a theory about the basic needs and wants shared by all individuals. He called them *stable preferences*, and he combined them with RCT to explain *all* choice behavior, not just buying and selling, as the result of economic calculations made by rational choosers. Then there was his conception of *human capital*—the knowledge and skills that individuals invest in acquiring. Becker's entire career was spent arguing that RCT does a great job explaining individual choice, not just market rationality, and does it better than the models of all other social and behavioral sciences. He couldn't have done any of the things that ultimately won him a Nobel Prize by treating RCT as just a summary of the laws of supply and demand.

There is a powerful reason Becker's original spin on RCT never took hold among other economists. Economic theory has had to treat RCT as a model of human choice that explains the rest of the economy. As we'll see, all contemporary macroeconomic theory would be unintelligible if we couldn't take RCT seriously as an explanation—not just a summary—of the laws of supply and demand.

Becker's rationale for RCT and Friedman's indifference (indeed, economic theory's indifference) to its idealization won't work. For one thing, it doesn't do justice to the way the theory is used in economics, not just to derive theorems about supply and demand but to explain why we can't turn these mathematical theorems into empirical facts. As we'll see, RCT is an idealization that economic theory employs to make sense of more idealizations. But even this fact about RCT doesn't answer the question of why economic theory can't do without it, no matter how idealized, unrealistic, or false it remains as a description of human choice.

Stylized Facts: The Best Economic Theory Can Do

The term *stylized fact* was invented by an economist, Nicholas Kaldor, to describe something that everyone in economics accepts as a fact but that can't be quantitatively described or measured. The label is now in wide use by economists (and elsewhere in the social sciences colonized by economic theory). Often a stylized fact stands out to the trained economist as obvious from a large body of empirical data. But it's also a fact that can't be sharpened into a precisely, formally, quantitatively, or explicitly expressed generalization about the data. This is what makes economic theory different from theory in the sciences. There, the data are the targets of explanation. In economic theory, by contrast, the explanatory targets are the stylized facts that economists informally agree stand out from the actual data. And that's a problem.

In the natural sciences, theory explains stylized facts on the way to explaining quantitative versions of these facts. That's how we know that an explanation of the stylized fact is on the right track—sooner or later it gives way to an explanation of the quantitative version of the stylized fact. This doesn't happen in economics. We need to see why and how this fact keeps RCT in its place at the center of economic theory. The best economic theory can do is explain stylized facts—the purely qualitative patterns in the economy that cannot be refined. Because of this, the explanatory models given for these patterns can't be subjected to an iterative process that goes from data to model and back to data to refine the model further.

Downward-sloping demand curves are stylized facts. There are, of course, real demand and supply curves connecting actual data points. And the economist has a well-developed explanation for why the closest we can get to explaining them are the stylized facts of economic theory. But this exculpatory explanation is ironical, as we'll see. It relies on taking RCT seriously, not merely as a shorthand for market rationality, but as a psychological theory about how economic agents think through their choices—just what their late nineteenth- and twentieth-century predecessors wanted to avoid.

If we could produce reliable, replicable demand and supply curves that remained relatively constant over time, and if RCT could, by itself or combined with other resources, enable us to derive the detailed shapes of these curves and the data points they connect, then RCT would be able to do more than explain stylized facts. The trouble, as noted above, is that it's pretty well impossible to collect data on real demand curves for real goods and prices that retain anything like sufficient stability over time to provide a test of RCT's predictions or a database that might help us improve them. There are just too many other changing variables in the real economy and in buyers' heads that make one day's or one hour's or one minute's demand curve different from the next.

These facts about the continual variability of demand curves are easy for the economist to explain. But to do it they have to take RCT seriously as a theory of what considerations people weigh in making choices when prices change. Let's see how.

The downward-sloping demand curve derived from RCT is the result of at least two component forces acting on people's choices. There is an *income effect* and a *substitution effect* that add together to determine how rational choice responds to price changes. The rational consumer's total purchasing power changes when the price of a good changes. At a higher price, the consumer has less left for other purchases; at a lower price, the consumer has more. That's the income effect. A simple example: as the price of beef increases, fixed-income consumers can continue to purchase the same quantity of beef only by reducing purchases of other commodities. But they can substitute other commodities for beef. They can buy pork or poultry or prawns. Each of these factors will affect their demand for beef.

RCT implies that actual choices that produce a consumer's demand curve can be decomposed into income and substitution effects. We can even draw diagrams that show how demand curves are the result of income and substitution effects. But these pictures just record more stylized facts. In a market with many substitutes and complements and many price changes among them, the shape of actual demand curves will be changing continually. So will the income and substitution effect curves that combine to form the demand curve. To use them to get to actual demand curves from RCT, you'd have to factor in a lot of information about available substitutes and their prices, along with information about how the consumer's income is changing.

Although there aren't stable long-term data to fill out the details, economists treat the income and substitution effects as parts of the "explanation"—the derivation of more stylized facts about demand, including stylized facts about stylized facts, in this case that sometimes the stylized fact of downward-sloping demand breaks down. Most demand curves slope downward. Not all. There are inferior goods—cheap stuff people buy less of as their incomes rise. The richer people are, the less instant ramen they buy (unless they really love ramen), even when its price declines. Then there are conspicuous consumption goods—the demand for Ferraris may rise as their prices rise to match those of Lamborghinis. There are markets where demand suddenly goes to zero at any price, like the market for slide rules with the advent of handheld calculators. In other markets, demand is constant no matter the price—think of bottled water in flooded areas with contaminated sewer systems. Economists explain all these stylized cases the same way they explain the standard stylized case of downward-sloping demand.

Each case is the result of income effects and substitution effects in consumers' calculations, or that's what it looks like: People who want to signal their wealth will buy more Ferraris as Ferraris get more expensive. Engineers will substitute calculators no matter how cheap slide rules are (until they become collectors' items in a completely different market, where they might turn out to be like Ferraris—collectibles that generate more demand the higher the price). On the other hand, people will pay any price for drinkable water if there is no substitute.

Economists seek to explain real cases by deriving the stylized facts from RCT. But they can't, don't, and won't go further to explain actual levels of demand or actual price movements. There isn't much point in doing so once the stylized fact is satisfactorily explained. The details can be left to market research, and besides, whatever market researchers discover won't have much use the next time reality presents a stylized fact in gory, non-stylized detail.

But here's the problem for economic theory's use of RCT to explain stylized facts. Without the ability to get to non-stylized quantitative versions of stylized facts, there isn't any independent, factual, empirical evidence that RCT explanations of stylized facts are correct. Without the iterative process of modeling, comparing a model's quantitative predictions with actual data to check, improving the model, and repeating the process, there is no reason to think the original model really explains or is even part of the story that explains a stylized fact to begin with. Unless, of course, you have reasons independent of factual evidence to accept the RCT model of choice. Do economic theorists have any?

RCT Will Always Be with Us

Outsiders to economics are left with the mystery of its continued attachment—nay, fixation—on rational choice. The reasons to embrace an idealized model that operate in the rest of science just can't get a grip on economics. And yet, economists insist their interest in making use of the model is purely scientific. But when a self-consciously experimental research program is undertaken to improve or replace RCT—see behavioral economics—the results have almost no impact on economic theory, in spite of the interest they generate among the general public. We've seen why. There just isn't any prospect of a cycle of experiments improving models and models suggesting new experiments that will amend RCT by reducing its idealizations and increasing its realism. At the same time, economists are hostile to alternative models—decision-heuristic rules of thumb—that provide wholesale substitutes for the provisions of rational choice theory.

As an explanation of why economists are so attached to RCT, the appeal of market rationality has some attraction. Economists really are much more interested in markets than in individuals. So why not drop RCT altogether? Economists could save themselves a lot of grief, and unburden themselves of *Homo economicus* forever, if they could bring themselves to abandon RCT.

In the next chapter we'll see why they can't, wont, and maybe shouldn't drop their commitment to rational choice theory. It's indispensable to an important insight that almost all economists share about how we *should* organize an economy. And that commitment to how things *ought* to be arranged has come to dominate their beliefs about how things actually *are* arranged.

3 Adam Smith Is Still Setting the Agenda for Economic Theory

According to most mainstream economists, economics got on the right track with Adam Smith in the 1700s. Smith made a remarkable claim: that the apparently unorganized, chaotic, anarchic, unplanned, and unpoliced conduct of self-interested economic agents—buyers and sellers—on every market results in outcomes that are beneficial to all. That wasn't the only thing he asserted, but it is the conjecture that economists hold to be his most significant insight.

I use the word *conjecture* here in the way mathematicians employ it. A conjecture, in mathematics, is a statement that seems to be true (at least for all the numbers we can try it on) but for which we lack a mathematically rigorous proof. Fermat's last theorem—that there is no number greater than two that satisfies the equation $a^n + b^n = c^n$—was a conjecture that worked for every number anyone tried, no matter how large. But it wasn't proved from the time Fermat conjectured it in 1637 until 1995. The job of the mathematician is to convert a conjecture into a theorem by giving a proof—a deductive derivation from secure postulates. The same goes for mathematical economists.

For almost a century a long line of the best economic theorists sought to turn Smith's conjecture into a proof. They finally succeeded, to their own satisfaction, in the 1950s. Meanwhile, during that century, the search for a proof of Smith's conjecture drove much of the development of economic theory.

Smith expressed his conjecture several times in *The Wealth of Nations*, but most clearly when he wrote that

> it is not from the benevolence of the butcher, the brewer, or the baker, that we expect our dinner, but from their regard to their own interest. We address ourselves, not to their humanity but to their self-love, and never talk to them of our necessities but of their advantages. . . .
>
> Every individual necessarily labors to render the annual revenue of the society as great as he can. He generally indeed neither intends to promote the public interest,

nor knows how much he is promoting it. . . . He intends only his own gain, and he is in this, as in many other cases, led by an invisible hand to promote an end which was no part of his intention. . . . *By pursuing his own interest he frequently promotes that of the society more effectually than when he really intends to promote it.*[1]

According to Smith, the invisible hand of the competitive market inevitably drives self-seeking individuals to an outcome that promotes the interests of society.

In a much less well-known passage of a prior work, *The Theory of Moral Sentiments*, Smith made another claim about the advantages conferred by the competitive market's invisible hand:

> The rich only select from the heap what is most precious and agreeable. They consume little more than the poor, and in spite of their natural selfishness and rapacity, though they mean only their own conveniency, though the sole end which they propose from the labours of all the thousands whom they employ, be the gratification of their own vain and insatiable desires, they divide with the poor the produce of all their improvements. They are led by an *invisible hand* to make nearly the same distribution of the necessaries of life, which would have been made, had the earth been divided into equal portions among all its inhabitants, and thus without intending it, without knowing it, advance the interest of the society.[2]

Combined, these two claims constitute a strong conjecture: that the competitive market produces both efficiency and equality—the largest quantity of commodities that people actually want, and a distribution of them that does not radically depart from equal portions. Both, Smith held, were in society's interests.

In their attempts to convert Smith's conjecture into a proof, mathematical economists had to surrender Smith's claim about equality, or equity, as economists now label it. Nineteenth-century mathematical economics may not have wanted to give up the claim that competitive markets had this morally desirable feature, but they had to, largely for "scientific" reasons. By the middle of the twentieth century almost all economists came to agree that economics itself could identify only efficiencies and inefficiencies, not inequities that concerns for justice might require us to eliminate. That matter was left to politicians and political philosophers.

To see why, and how, the search for a proof of Smith's conjecture had to change it to stand a chance at success, we have to do a little history of economic theory.

Self-Interest and Introspective Psychology

The first axiom in a proof that the competitive market is beneficent is Smith's oft-repeated assertion that individuals pursue their own interests, that they intend only their own gain, and that they are driven by their "self-love."

Nineteenth-century economists rephrased this claim into an assumption that individuals seek their own pleasure, happiness, or satisfaction, and then formalized it into the notion that people are rational utility maximizers. *Utility* is a label for the *psychological stuff* variously described as the subjective feeling of pleasure, satisfaction, or happiness, or sometimes more broadly as welfare or well-being, that is universally desired for itself and not as a means to some other psychological state. At least one of the founding fathers of utility theory supposed this insight was an important contribution to psychology. Francis Ysidro Edgeworth entitled his deeply influential book on the subject *Mathematical Psychics*.

Like Smith, nineteenth-century English economists were also moral philosophers. In fact they were explicitly utilitarian moral philosophers. They endorsed the maximization of happiness, measured in utility, as a moral ideal and objective. Actions and institutions were to be assessed for moral goodness by whether they enhanced the total quantity of happiness of those affected. Although utilitarians disagreed about the nature, gradations, and sources of happiness, they continued to hold that the morally best outcome among alternatives was the one that maximized total utility. The attraction of Smith's conjecture to these economists–cum–moral philosophers was obvious. If he was right, the competitive market would turn out to be morally desirable due to its utility-maximizing outcome.

The reformulation of Smith's postulate that individuals are self-interested into the claim that they are utility maximisers made his starting point more explicit. If individual amounts of utility could be added up to determine under what circumstances the total was maximized, Smith's claim could be converted into a much more precise hypothesis about the market: when operating properly, it leads individual self-interested utility maximizers to also maximize the total utility in society. Economists could then try to prove that was true.

The obvious trouble with this way of formulating Smith's conjecture is that it conflicts with what introspection tells us about the psychological characteristics of human happiness, pleasure, and the satisfaction derived from consuming market goods and services. Our own levels of utility don't come in measurable amounts.

For a start, the quantity of pleasure, happiness, or whatever it is we derive from consuming goods varies too greatly and too rapidly to be a stable quantity we can measure and then use to make predictions. How much satisfaction we secure from anything depends on the availability of complements and substitutes, which itself fluctuates over time. Squeeze ten grams of mustard in my mouth and I will suffer discomfort. Squeeze it in along with a bite of hot dog and I will secure some pleasure. How much? That will depend on how many bites of hot dog and mustard I have already taken. If I prefer ketchup to mustard on my hot dog and it's available, then the pleasure of hot dog and mustard may be less than it would be if ketchup were not an available alternative. Most of us can rank feelings of pleasure or satisfaction on a scale of greater than and less than: contemplate the enjoyment of differing wines or ice cream flavors. But even if our imaginations and memories give us reliable access to prospective or previous pleasures, we can't isolate them in discrete amounts—how much utility they provide. *Pleasure, happiness,* and *welfare* are mass nouns, like *water* and *snow*. They do not exist in natural (cardinal) units (like apples: one apple, two apples) but in amounts for which we need to construct measures (such as grams or ounces). We can coin a label, such as *util*, but that doesn't make it measure anything real. We can be confident we have a reliable unit of measure for a substance only when it enables us to establish testable regularities about the substance's behavior. Thermometers tamed heat into temperatures—units we could measure. We knew they were reliable units because the thermometer enabled chemists to discover the gas laws. Only scientific success vindicates units of measurement. There has never been any such success for utils.

It took the nineteenth-century "marginalist" economists a few decades of thinking about measurement to recognize that *utility* doesn't name a psychological quantity that individuals can seek to maximize. That doesn't mean that people don't act in their self-interest or seek their own pleasure above all things. It only means that utility and its units don't figure in psychological explanations of what people do.

Utilitarianism demands that we add up the utilities that each individual agent secures from each possible distribution of commodities and then compare the sums for total size. The trouble is, first, that we can't get inside anyone else's head. Second, if their utility doesn't come in units, we can't add up several people's utility. Anyway, our inability to access the psychological states of others makes it impossible for us to compare one person's level of happiness, satisfaction, or pleasure to anyone else's, and still more so to find a way to add up each individual's amount of happiness under diverse circumstances and compare the totals.

Nineteenth-century economists employed utilities in their theories to explain why people act rationally—that is, why they make choices in the way RCT describes. Each of the four axioms about preferences stated at the beginning of chapter 2 follows from the hypothesis that people maximize utility. Think about it: if you were a utility maximizer omniscient about the alternatives you face, you'd choose in strict accordance with RCT (except perhaps when it came to nonsatiation—sometimes enough is enough and we don't want any more).

Introspection about satisfaction, pleasure, and happiness teaches us at least one more important lesson the economist thought was critical for explaining exchange.

Each of us recognizes by introspection that we secure a diminishing amount of pleasure or satisfaction from each additional (what economists call *marginal*) unit of a good we consume. None of us can directly establish that this is true of any of the rest of humanity. But we're pretty confident about it because people's choice behavior seems to be driven by the same diminishing marginal satisfaction that we experience ourselves.

The introspective feelings we all experience were formalized into a "psychological law" of diminishing marginal utility. Put together with RCT, this law explains why demand curves take their shape. As quantity consumed increases, marginal utility decreases, and so willingness to pay decreases.

But there is a problem with establishing diminishing marginal utility as a psychological law: there is no way to measure it in behavior. Nineteenth-century marginalist economists recognized early on that decreasing willingness to pay money for the marginal unit of consumption was not reliable evidence of a general psychological law of diminishing marginal utility. Willingness to pay is a matter of how much money an individual has. There is, in the words of the economist, an *endowment effect*. Additionally, money is itself supposed to be subject to diminishing marginal utility. This makes circular its use to measure the very psychological process that explains people's willingness to spend or work for money.

Nineteenth-century economists were increasingly embarrassed by the psychological theory—the *Mathematical Psychics*—they had saddled themselves with. It wasn't just that no one could contrive an instrument for measuring subjective individual utilities or add them up. Thought experiments (and real ones in psychophysics, or perceptual psychology) made it increasingly obvious that there were no such well-behaved psychological units or quantities that could be measured. Economics was saddled with a psychological theory that was never going to get any better evidentially, predictively, or explanatorily.

Normally, when this kind of thing happens in science, finding no units for a substance or quantity that link it to something else we can measure, scientists conclude that there is no such thing. Two famous examples are impetus and caloric. The first of these was defined in pre-Newtonian physics as the force that is required for an object to have nonzero velocity. Because force is required for nonzero acceleration, and not for nonzero velocity, there is no such thing as impetus. Despite a thousand years of looking, no regularity about the motion of objects could be found to measure this force. Physics gave it up. *Caloric* named the conserved fluid substance that flows between objects and contains heat. Again, no units for caloric were ever found to measure it directly or indirectly. No wonder, since heat is not a fluid, nor a substance at all; it's molecular motion.

The difference between these concepts and utility is that economists, along with everyone else, knew that pleasure existed. That made it hard to give up the word *utility* as a label for these feelings, even as they recognized that the label was as scientifically intractable a notion as impetus or caloric. There is no such thing as utility, and economists needed to avoid talking about it altogether. Many of these economists began to want to reconfigure their discipline into one that was separate from psychology. "Psychologism" was already, in the nineteenth century, a term of abuse employed by others, especially institutional and historical economists, to condemn the mathematically expressed theories of those who sought to convert Smith's conjecture into a proof.

Economics had to expunge any commitment to psychological causes and psychological explanations from its theory. But it had to throw out the bath water without the baby. It needed to save as much of the explanatory theory as could be cleansed of psychological commitments, but not so much that it would have to give up the quest to convert Smith's conjecture into a proof.

Indifference, Not Utility; Efficiency, Not Equity

Enter Vilfredo Pareto (1848–1923). Pareto was a behaviorist avant la lettre. A generation later John B. Watson's and later B. F. Skinner's behaviorist research programs swept through psychology and begin to spill over into other social sciences. By the time economists noticed that they, too, were behaviorists, the payoff that behaviorism promised had been fully discounted in their discipline's embrace of Pareto's insight.[3] Once Pareto's take on utility became orthodoxy, economic theory was finally ready to come as close as it could to proving Smith's conjecture about competitive markets.

Behaviorism in economics is, functionally, the injunction to ignore the causes of choice behavior and take the choosing behavior itself as the starting point. We can't measure how much happiness, pleasure, or utility an agent gets from a commodity, but we can tell with perfect confidence whether the agent prefers one good over another by offering the agent a choice and observing which they select. We can establish what economists call an agent's *ordinal preferences*. We offer alternative choices and seeing how the agent ranks—orders—them. Ordinal preferences are all we need for economics, so we can forgo any commitment to utility or other psychological quantities or qualities we might believe to be driving preference rankings.

Instead of theorizing about causes of choice, Pareto advised that we simply focus on actual choices. Assume that the chooser strictly obeys RCT. Offer the chooser two options. The agent will choose one or the other or will be indifferent (rank them as equal). The choice the agent makes is observable. Later economists called this *revealed preference*, but that's misleading, since economics is not concerned with what the choice reveals about its mental causes—the agent's subjective tastes. Offer an RCT-obeying agent a large number of choices between different amounts of a commodity, pairs of commodities, or other alternatives and you can build a spreadsheet of the agent's preferences. This ordered list is the rational agent's revealed preference. "Agents are rational" turns out to mean only that their actual choices obey all four of the conditions of RCT, regardless of what psychologically drives these choices.[4]

Though economics turned its back on utility as the psychological determinant of preference, the field continued to employ the expression *ordinal utility* as a synonym for preference rankings or orderings. Rational agents are said to maximize ordinal utility. But no one was misled into thinking that there is such a thing as ordinal utility—something in people's heads that comes in amounts that exist only on a more than/less than scale, instead of a cardinal scale of how much more than or less than. *Utility* is a word that economists continued to use without taking it seriously, as we'll see in chapter 4.

No economist actually experimented with offering people a comprehensive set of choices and constructing their revealed preferences. It's easy to see why they didn't bother. People's choices don't stay the same over time; they vary with changes in taste and with substitutes and complements, and there are just too many commodities over which people make choices for the data to be manageable. In addition, economists didn't need to do experiments. They were all pretty confident in what the resulting data would look like. Introspection already told them that their choices would be driven by the feelings of diminishing satisfaction we all experience: the more of anything you consume, the less satisfaction you get from the unit most recently consumed (the *marginal unit*). Everyone reports that they experience diminishing marginal utility. Of course, after Pareto, economists couldn't allow themselves to make use of their subjective experience as evidence. It was psychology, and thus off-limits. But Pareto figured out a way to derive the market rationality that economics is supposed to be interested in from diminishing marginal utility without ever mentioning utility. We'll see how in the next section.

But while discarding a deeply flawed psychological theory of choice, Pareto's followers also had to abandon Smith's claim that the market made everyone about equally well-off and significantly weaken his conjecture that the invisible hand improved society as a whole. Once we give up the maximization of utility, either individual or total, as unintelligible because it is unmeasurable, we also have to give up the maximization of total utility as a societal goal that morality imposes on us. Pareto saw this and recognized that it put an end to any attempt to prove Smith's conjecture as the utilitarians conceived of it. If only revealed preferences exist, what would it even mean to ask whether the market maximizes total revealed preferences?

The preferences of what? Of society? Of the sum of economic agents taken together? The idea makes no sense.

Pareto realized that we might be able to provide a different, weaker, less morally attractive conclusion than Smith's conjecture. But at least it would be one in the spirit of Smith and his utilitarian followers. And it would be a "result" we could actually prove mathematically. What Pareto proposed as a substitute for the utilitarian objective of maximizing pleasure, happiness, and welfare has borne his name ever since: *Pareto optimality*, an observable, empirically decidable replacement for the measure of maximizing total utility.

This is how it works: Trade in the market may not make people measurably better off if utility is what is measured. But any particular trade makes both parties better off, at least in principle. Otherwise they wouldn't trade, would they? Each party secures a more preferred alternative to their initial starting point—their initial endowment—by trading. Therefore, starting from an initial distribution before trade, free exchange should make at least some people better off while making no one worse off. Assuming people are rational (according to the four assumptions of RCT), fully informed, and not coerced, then once everyone has had a chance to trade with everyone else, the distribution that results after all trades have been made should be Pareto optimal: no one can be made better off by any further redistribution (by trade or any other means) without someone being made worse off. And we can tell when a distribution is Pareto optimal by switching things between parties and seeing whether people complain.

Pareto optimality is the closest economists think we can get, empirically, to measuring the welfare benefits of the market—or of anything else, for that matter. Economists from Pareto onward substituted Pareto optimality for welfare maximization in Smith's conjecture. Then in the early twentieth century they set out to prove that trade in a perfectly competitive market would result in a final distribution—a *general equilibrium*—where there would be no more trading, and it would be a Pareto optimal distribution.

To prove: Every general equilibrium resulting from a perfectly competitive market is Pareto optimal.

This statement is famously called the first theorem of welfare economics—the subdiscipline of economics that explores how individual welfare, well-being, happiness, and satisfaction are affected by trade.[5]

The first thing to notice about this refinement of Smith's conjecture is that it changes the ambitions of economics as a discipline. From Smith to Pareto, economics was a compartment of moral philosophy dedicated to making everyone—humanity, society—better off. After Pareto, economists decide that their discipline could or should be a value-neutral empirical science.

The shift from seeking to prove that the competitive market maximizes happiness, pleasure, or welfare to seeking to prove that it produces a Pareto optimal general equilibrium means that we can no longer prove that the competitive market attains the utilitarian's morally endorsed outcome—maximizing total utility. The most that we can aim to prove is its economic *efficiency*: that exchange on the market redistributes tradable goods in a way that doesn't waste any of them. Given the inputs to production of goods and services, agents' preference rankings among these inputs, and people's rational, self-interested, pleasure-seeking behavior, the final result of trading in a perfectly competitive market won't leave anything desirable undistributed and won't leave anyone without a wanted item they are able to buy at the market price.

Proving the efficiency of the competitive market is something, but it's not everything Smith hoped for. In particular, proving efficiency doesn't prove that the final efficient distribution is morally desirable. Recall Smith's second conjecture, that exchange minimizes the differences between rich and poor in regard to their welfare. Smith at least implicitly endorsed the moral value of something approaching equality—some sort of equity as the outcome of exchange. The rich, he wrote, "are led by an *invisible hand* to make nearly the same distribution of the necessaries of life, which would have been made, had the earth been divided into equal portions among all its inhabitants, and thus without intending it, without knowing it, advance the interest of the society."[6] But it's clear that a Pareto optimal distribution need not be an equal, fair, or just distribution. Think about it: A distribution of goods and services that results in one person having everything and any number of others having nothing is Pareto optimal. Those with nothing cannot be made better off without making the owner of everything at least a little worse off. A distribution that deprives some simply to enrich others could produce a new distribution that is also Pareto optimal. A random distribution can be Pareto optimal as well.

Can we argue that being Pareto optimal is at least one rather weak and uncontroversial feature of a morally permissible or equitable outcome?

"Waste not, want not" seems a morally worthy ideal, all things being equal. But it's nowhere near what most people's moral commitments require.

There is one thing worth noting about Pareto optimality as an objective measure of evaluation. All parties to an inquiry about whether a distribution is Pareto optimal can agree on what counts as Pareto optimality. Allowing for experimental or observational uncertainty, they can agree on whether any given distribution meets the test of Pareto optimality. But this kind of agreement on empirical facts is never enough to settle any moral, ethical, or evaluative dispute.

Once economics substituted Pareto optimality for maximal total utility it could try to convert Smith's conjecture into a proof. But it had to give up the hope of ever endorsing the outcome of market competition as morally ideal. Since no empirical science can underwrite any outcome as morally right, this was a loss economists who wanted to be scientists could contemplate with equanimity.

Since the 1930s economists have almost universally accepted as their definition of the field one that makes efficiency in Pareto's sense paramount. The definition was framed, canonically, by Lionel Robbins: "Economics is a science which studies human behavior as a relationship between ends and scarce means which have alternative uses." Robbins made clear that economics was no longer about ends, only about means, and everyone took it to heart: "Economics is entirely neutral between ends. . . . In so far as *any* end is dependent on scarce means, it is germane to the preoccupations of the economist."[7]

Turning Smith's Conjecture into a Proof

Pareto needed something measurable that would do the work of diminishing marginal utility to derive downward-sloping demand curves, since there was no such thing as utility to diminish marginally. He came up with the concept of an *indifference curve*. It turned out to be enough to prove the strongest version of Smith's conjecture that could be stated without utilities, the first theorem of welfare economics. We'll sketch the proof using some classic diagrams from economic theory.

Start with a rational agent—one who acts strictly in accordance with RCT. Offer the agent choices between any two commodities or bundles of two or more commodities, say apples and bananas. Offer a choice between a package of [2 apples, 3 bananas] and a package of [2 apples, 4 bananas]. The agent will have a preference because of completeness of preference rankings (the first assumption of RCT). The agent will pick the second because of nonsatiation (the second assumption): people always want more of anything they want at all. But there will be some bundles of apples and bananas people will like equally well: say [3 apples, 2 bananas] and [2 apples, 3 bananas]. The agent will be indifferent between these two bundles. Now make a list of all the bundles that the agent can't choose between. Then graph the data and connect the dots. Let's say the agent is indifferent to [8 apples, 0 bananas], [6 apples, 1 banana], [3 apples, 3 bananas], [0 apples, 5 bananas]. The curve connecting these points is an *indifference curve*. There will be many such curves for the different combinations of commodities the agent is indifferent to, none of them intersecting (if the agent is rational). All of them will be convex from the origin, as in figure 3.1.

Convex indifference curves are all we need to prove the first theorem and derive a weakened version of Smith's conjecture as a mathematical consequence of RCT. But before we see how this proof works, why should we assume that all indifference curves of all rational consumers are convex? Well, here is an argument that might convince you and me, but that no economist can use.

Assuming that other people are like ourselves, they'll derive utility—satisfaction—from consumption and their consumption will be subject to

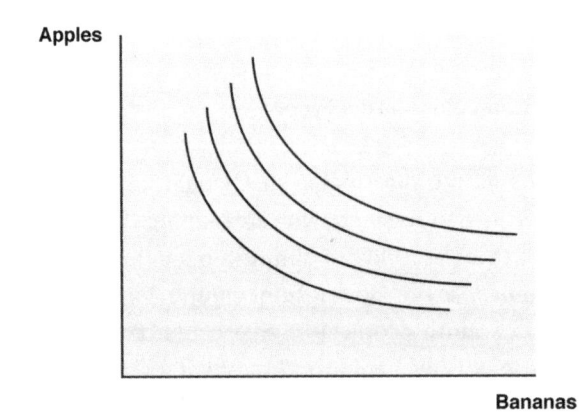

Figure 3.1
Four of the infinitely many indifference curves for some agent and two commodities—in this case, apples and bananas.

decreasing marginal utility. Suppose people like apples and bananas equally well. As the number of apples consumed increases, people get less and less satisfaction from the last one consumed. The same goes for bananas. Therefore, they will be indifferent between bundles of apples and bananas where the marginal utility of one more apple and one fewer banana is equal to the marginal utility of one more banana and one fewer apple. The indifference curve resulting from downward-sloping marginal utility will always be convex.

Arguing from your own introspection isn't a scientifically admissible reason to expect that all indifference curves are convex from the origin, but it is a powerful subjective motivation for thinking so. If you try the experiment, offering options to people and then constructing real indifference curves from their choices, the curves will be more or less convex for most people, most of the time. But they'll have gaps where you lack data points and violations where people didn't pay attention or were confused or couldn't decide. Even if you use the same subjects two days in a row the data might differ. It's enough to make everyone confident that the space of bundles of any two commodities is dense with convex indifference curves.

Some economists felt a need to explain why indifference curves are convex without appealing to forbidden psychology. They substituted the nearest observationally testable principle they could think of: a "law" of *diminishing marginal rates of substitution*, DMRS, between commodities. The more of any one commodity you consume, the smaller the amount of any other commodity you'd be willing to give up to consume an additional unit of the first commodity. If the law of DMRS is in force among all pairs of commodities, then all indifference curves are convex. But why suppose it holds between, say, fishhooks and textbooks? There isn't much evidence in favor of this particular instance of DMRS. We're confident of the relationship between butter and margarine. But that's because introspection tells us they're substitutes and that we'll get less satisfaction from consuming more of one as we consume more of the other. We can't use introspection in economics. Evidence for DMRS emerges only when it's tested by varying close, real substitutes in bundles of commodities that the consumer is indifferent between. But how can we tell whether two things are really close substitutes scientifically? Only by using DMRS. Any experiment that will vindicate DMRS for different combinations of goods is also evidence that their indifference curves are convex. But that's where we came in, looking for the reason that indifference curves are convex. DMRS just states this stylized fact in other words; it doesn't explain why. The fact is economic theory just assumes indifference curves are convex because it must in order to prove Smith's conjecture without using diminishing marginal utility.

Given convex indifference curves, it is easy to use a cute diagram to show how trade invariably produces a Pareto optimal general equilibrium, at which point all trade ends. It's a graph of the simplest case of an economy with two traders and two goods. Generalizing the result to an economy of n traders and m goods, where n and m are large numbers, is another matter. We'll discuss it briefly later. The graphic device used to do this has been around since Edgeworth used it in *Mathematical Psychics*, taking psychological utilities seriously. It was Pareto who realized it would work for preference and indifference without any need for utilities.

The diagram represents the situation of two traders of two commodities. Able and Baker have some initial endowment of apples and bananas, and they each have indifference curves for various combinations of these two commodities that look like the ones in figure 3.1.

If we combine graphs of Able's and Baker's sets of indifference curves in a certain way, we'll reveal their trading opportunities. Draw a box or rectangle. Then add Able's indifference curves convex from the usual (0,0) origin at the lower left corner of the box, and Baker's indifference curves convex from the upper right corner. The result will be two sets of indifference curves, one set for Able, one for Baker, each curve facing another one, carving out a set of lens-like spaces. Of course, we can't draw all the indifference curves. There are indefinitely many of them if the two commodities are infinitely divisible. So there will be indefinitely many lenses composed of one indifference curve from each agent. Figure 3.2 contains four lenses composed of eight indifference curves—four of Able's and four of Baker's.

Now, any point inside the square represents a particular package of some number of apples and some number of bananas for each of Able and Baker where the total quantity of apples is represented by the length of the x-axis and the total quantity of bananas equals the length of the y-axis.

Pick any starting point inside the rectangle. It will give us Able's and Baker's initial endowments of apples and bananas. Since every point in the box is the intersection of two indifference curves, one for each of the traders, the starting point we picked lies at the intersection of one of Able's indifference curves and one of Baker's. These two curves outline a lens. Every point

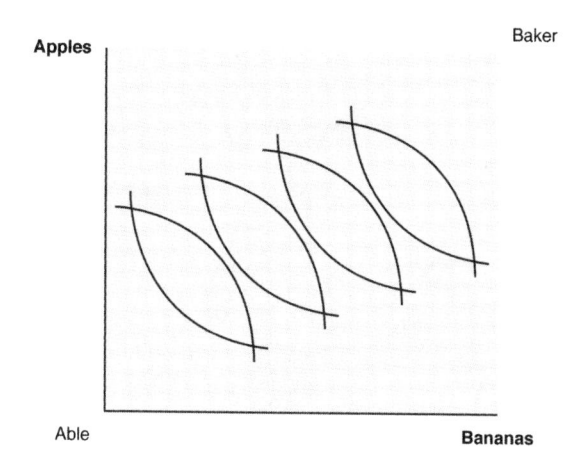

[Figure 3.2: Edgeworth box diagram with Able at bottom-left origin and Baker at top-right, Apples on vertical axis, Bananas on horizontal axis, showing intersecting indifference curves forming lenses.]

Figure 3.2

inside that lens is a different distribution of the same total number of apples and bananas the traders started with. But every one of them is preferred by both Able and Baker to the starting distribution. In figure 3.3 the lens is the shaded region, or the lens of Pareto improvements, for both Able and Baker.

What will they do? If they are rational bargainers, they will pick a new point, any point, inside the lens, and bargain to exchange apples and bananas until they reach that point. Why? Because such a trade is a Pareto improvement on the previous distribution since it takes each trader to a point on a higher indifference curve. This second point is also intersected by two indifference curves that draw a new lens, smaller than the first and completely inside it. The new lens contains points, attainable through another round of trading, at which both agents will be even better off. Repeating the exchange between Able and Baker produces an even smaller lens. The process can continue until the lens disappears at a point where two indifference curves are tangent to one another. There is no lens left.

When the two agents find themselves with quantities of apples and bananas represented by the point at which their two indifference curves have shrunk the lens to zero, they will cease trading with one another. At that point the distribution of apples and bananas between them will be both Pareto superior to their initial endowments and Pareto optimal. Able and Baker won't both be able to do better.

Economists call this state, at which no further trades between the parties take place, an *equilibrium*, suggesting that economic exchange has come to

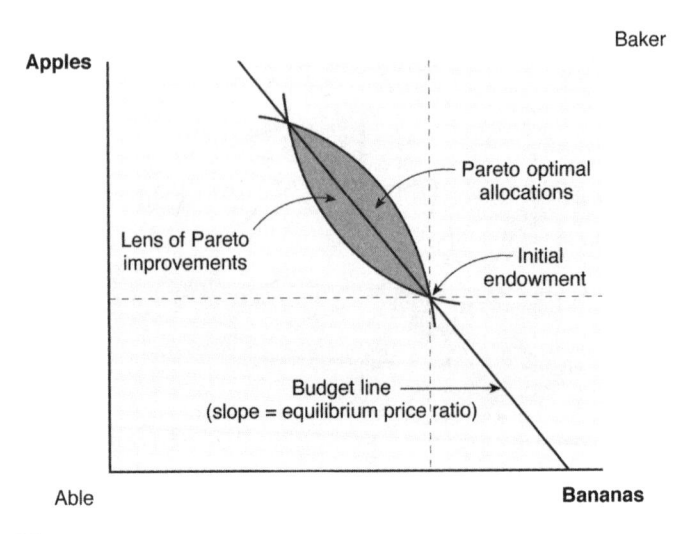

Figure 3.3

Labels in figure: Apples; Baker; Pareto optimal allocations; Lens of Pareto improvements; Initial endowment; Budget line (slope = equilibrium price ratio); Able; Bananas

rest. If apples and bananas are the only two goods traded in the economy, the resulting equilibrium is a *general equilibrium* where the whole economy is at rest. At this point there is no trading at all, and both parties recognize there will be no Pareto improvements from trade. The distribution is Pareto optimal.

It's a relatively simple matter to extend the Edgeworth box analysis to show that producers or firms seeking to maximize profits will trade inputs to their productive processes in ways that result in Pareto-optimizing exchanges. All we need is to assume that their production possibility curves, which measure the intersubstitutability of productive inputs, are convex, like consumers' indifference curves. The goal of maximizing revenue making either apple cake or banana bread will lead producers to exchange apple and banana inputs until all the gains from trade have disappeared. We have good reason to think that production possibility curves will be convex to the origin if the marginal productivity of inputs declines. (Unlike marginal utility, this is something we can measure.)

To convert Smith's conjecture into a theorem, the graphical illustration of the inevitability of a Pareto optimal equilibrium between two rational traders needs to be expressed as a mathematical proof and proved for more than two traders and more than two commodities. That requires generalizing the Edgeworth box to n traders and m goods, where n and m are any positive numbers, which requires the application of results from topology unknown among mathematical economists before the mid-twentieth century. In 1954 Kenneth Arrow and Gerard Debreu provided the theorem independently (for which they were much later awarded the Nobel Prize in economics, once it had been established).

Economic theory points to this result to show how the real economy works when it's working well, in developed capitalist economies, and to show how the real economy needs to be changed when and where it doesn't. It drives the mainstream economist's commitment to free market economies, continues the invocation of Adam Smith as economic theory's founding father, and constitutes the pedagogical core of graduate education in economics and the framework of almost all research in economic theory. We'll find ourselves referring back to Arrow and Debreu's proof several times in the chapters to follow.

Economic theorists recognize the theorem's limitations even as the description of a theoretical possibility, let alone a real economy. But that does nothing to reduce their devotion to it as a prescription of how to achieve something approximating Smith's "interests of the whole society."[8] The proof has significant limitations, important qualifications, and initially unnoticed exceptions that are themselves matters of complicated theoretical reasoning, all of which can be appreciated only once you understand the proof fully enough to admire its achievement.

Even before the completion of the modern version of the proof—that there is a general equilibrium and it is always Pareto optimal—theorists were conducting thought experiments about goods and circumstances that would forestall attainment of the general equilibrium of Pareto optimality. Once the proof became canonical, the prizes for dreaming up counterexamples and conditions needed to exclude them became a competition for prestige among theorists. We will explore some of these ingenious exceptions—goods whose equilibrium supply is never Pareto optimal—in later chapters.

Even as they developed the proof, theorists recognized that applying it to the real world required idealizing, approximating, and sometimes squinting to see how things smoothed out in the ways perfect competition requires. The list below, of assumptions of the model, is familiar to economists but by no means complete. The assumptions deserve some elaboration.

1. Economic agents are rational preference maximizers. (Enough has already been said about this assumption.)

2. All commodities traded are infinitely divisible. This is a largely mathematical requirement that assures the curves and the mathematical functions they depict are continuous and smooth, without gaps or breaks. For a handful of apples and bananas, and in the absence of a knife, this is obviously a significant idealization, but for a large economy producing, say, eight million (indivisible) cars it's pretty harmless.
3. The number of traders (buyers and sellers) is very large—so large that no one can set prices. Every agent is a *price taker*. If the number of traders in any commodity is small, then a few of them could collude, refusing to trade at the Pareto optimal price, which would force others to take that price, resulting in a Pareto nonoptimal outcome: prices at which agents have too much or not enough of a commodity. If the number of buyers and sellers for any good is very large, then no one can corner the market, no one can be a *price setter*. Anyone trying to set their own price will be undercut by someone else ready to trade at the market price. In the Edgeworth box above, we implicitly assumed that Able and Baker were price takers, even though there were only two of them and each could withhold their goods from trade or perhaps even bully the other party into a trade.
4. The returns to scale are constant in the production of all commodities. Companies cannot become more efficient simply by increasing the size of their work forces, factories, or inputs.
5. All traders have access to all the information they need about commodities and other traders' preferences among them. In our Edgeworth box, Able and Baker couldn't hide information that might affect either party's offers.
6. The general equilibrium the proof guarantees is forever. All trades have been agreed to, not just for goods on hand, but for all future deliveries of these goods and all other goods to be made in the future. This *futures market* is one in which traders can buy and sell commodities not just for immediate consumption but for delivery to all places at all times in the future. A complete futures market enables individuals to rationally plan future consumption, production, investment, disinvestment, and more.

Even this list of assumptions is incomplete, as we'll see in later chapters. Some of these assumptions seem to be harmless, such as the infinite divisibility of commodities. But others betray serious problems (recognized and acknowledged by economists among themselves) in any attempt to use the model of perfect competition to understand the real economy. Think about how an indefinitely large number of traders would ever find their way to a general equilibrium even if they knew it had to exist for some set of prices. Two traders might repeatedly bargain with one another to find the final trade at the tangency of their highest indifference curves. But how could two hundred traders, or two million, do so, seeking to exchange a very large number of goods now and forever? How would they ever find a way to a set of prices that would clear all markets? Getting to the general equilibrium produced by a perfectly competitive market was a process no one could describe.

Three Cheers for the First Theorem of Welfare Economics

Now we know why RCT is indispensable to economic theory. The real reason economics will never give it up is simple: it's the key to converting Smith's conjecture about the invisible hand into a proof. And the centrality to economics of the theorem that the perfectly competitive market reaches a Pareto optimal general equilibrium cannot be overstated. As we'll see, almost everything else in economic theory is driven by the theorem.

The way economists employ the theorem is complicated. They use the model of perfect competition to explain the economy when it seems to work—when the model's idealized assumptions come close enough, in their view, to matching the realities of the market. But they also use the model when the economy doesn't work. When the real economy departs from anything that could be called Pareto optimality, the theorem becomes a template for intervening in the economy, tinkering to bring it more closely into line with the theorem. When the theorem seems to work, economic theory is descriptive, "positive," a "scientific theory" that helps us understand what is going on. And when the theorem fails to describe what's going on, the fault is in the economy, and so the economy has to be changed.

This is why economists won't—can't—give up RCT and the rest of the apparatus they use to prove the first theorem of welfare economics. When it works it's science, and when it doesn't work it's applied moral philosophy demanding that we achieve its normative end—Pareto optimality!

Economic theory is deeply anomalous. Its models, like others in the sciences, can be falsified. In fact, they're falsified all the time. That's what makes economists' track record of never improving their models' predictive accuracy so clear. But unlike other scientists, few economists have tried to give up on the early models and search out new ones that might do better in prediction. The kind of insulation from empirical refutation that economic theory has secured among economists is unique. It makes the discipline different from all other social and natural sciences.

This raises two obvious questions to the outsider: First, if economic theory is taken as a descriptive, explanatory endeavor, why should we think it works? Is there anything about its history that should give us confidence that its explanations are right, or even on the right track? Deriving "stylized facts" from idealized assumptions, and doing so after the fact, shouldn't give us much confidence in its post hoc explanations. The second question arises when the first theorem is treated as a moral imperative about how to arrange the economy: What's so special and desirable about Pareto optimality? Even if we admit that efficiency is a good thing, there are endless different distributive outcomes of a market economy that are all Pareto optimal, including many that are grossly inequitable, distributionally unjust, or morally objectionable. And there are others that are not Pareto optimal at all but that may nevertheless be morally preferable to some (or perhaps most) Pareto optimal outcomes.

4 Who Invited Calculus?

Economic theory is laden with mathematics, some of it quite formidable, and most of it formidable looking. The fact that some economic theory can be expressed in differential and integral calculus, particularly in forms familiar from physics, such as the one we'll discuss here, strongly suggests to the outsider that economics is much more like physics than it is like other social sciences. Much of the discipline's prestige has long derived from its expression in mathematical symbols. But the math should not be the barrier it is to understanding the theory. The mathematization of economic theory is not due to its absorption with money. In fact, as we'll see in chapter 5, it's hard to make money part of economic theory at all. Economic theory's mathematical "look" all starts with its focus on utility in the nineteenth century.

Mathematizing Utility

Edgeworth treated utility as a real, measurable psychological quantity that agents seek to maximize. This afforded him the opportunity to introduce differential calculus—the mathematics of minima and maxima—into economics. But even though utility later had to be excised from economic theory, calculus never left. That should make a lot of the mathematics of economic theory slightly puzzling. Economists ignored the puzzle. They liked the math.

Edgeworth and the other nineteenth-century economists treated utility as a psychological quantity that came in continuous amounts, not discrete units. And they were inordinately concerned with *marginal utility*—the amount of satisfaction produced by the last unit of a commodity consumed. Diminishing marginal utility is what gave them downward-sloping demand curves.

Marginal utility is what enabled marginalist economists (as they came to be called in retrospect) to invoke differential calculus. Once you define marginal utility as $\delta U/\delta x$, the partial derivative of utility with respect to one unit of commodity x, you're ready to rewrite everything you want to say about marginal utility using these impressive symbols. (Partial derivative just means that there may be variables other than x affecting utility that can change but are held constant while you figure out the number that gives the rate of change of U as one unit of x is added). But the partial differential definition of marginal utility doesn't get you the impressive math that economics shares with physics. That comes with a bit of calculus Edgeworth imported from eighteenth-century physics.

In the Edgeworth box that nineteenth-century economists used to illustrate the existence of a general equilibrium between two traders, each trader faced a *constrained maximization problem*—how to *maximize* utility subject to the *constraint* of a limited budget. The diagram showed graphically how both could simultaneously solve their problems. Edgeworth wanted to express the result mathematically. The solution to constrained maximization problems was already mathematically well understood in physics, and Edgeworth helped himself to the recipe physicists had been using to solve them. Over the next hundred years the recipe became a staple of mathematical economics, a source of endless textbook problems for econ students to solve, and a key tool in constructing macroeconomics and growth theory.

Here is how economic theory solves the economic agent's constrained maximization problem. Given an agent's total resources, or budget, and the prices of the commodities, the agent must determine how much of each to purchase to maximize utility. In a simple case of two commodities, x, and y, with prices $2 and $5, and $100 to spend, the budget equation is $100 = 2x + 5y$. Here it is, rewritten more conveniently for the physics technique that Edgeworth introduced:

Budget constraint: $0 = 2x + 5y - 100$

There is also an equation for how the agent's utility depends on the commodities the agent purchases: $U = f(x, y)$. Exactly what form can this equation take? Any mathematical form at all! The technique Edgeworth borrowed works with any mathematical function that can be subject to differentiation by methods of calculus. The way utility depends on commodities can satisfy any such equation, from the simplest to the most complicated. It

Who Invited Calculus?

could be $U = x+y$ or $U = \log xy$ or $U = x^2 3y^{1/2}$ or any of an indefinite number of formulas for one, two, three, a dozen, or any number of commodities mathematically related in any way, so long as the resulting mathematical equation can be subject to differentiation. Let's use one of the simplest for our example.

Utility equation: $U = xy$

where U is the quantity of utility derived from amounts of two commodities, x and y. To determine the values of x and y that will maximize U, given the agent's budgetary limitations, we use the following mathematical expression, called the *Lagrangian*, invented by physicists to solve their own constrained optimization problems for energy. Plug into the Lagrangian the two equations for the budget constraint and the utility function:

L = [The utility equation] – λ [The budget constraint equation].

In our case,

$L = [xy] - \lambda [2x + 5y - 100]$.

Now we take the first partial derivative of L with respect to each variable in the equation: x, y, and λ. There is a recipe in calculus textbooks for how to do this for almost any expression containing x and y. In this case, starting with the formula above, you are writing three expressions for how L is changed by a change in one of the variables, x, y, and λ, while holding the other two constant. The Greek δ in $\delta L/\delta x$ is used instead of dL/dx to remind us that we have to treat the variables y and λ as constants when we differentiate x, and so on.

$\delta L/\delta x = y - 2\lambda$ (because differentiating L with respect to x gives 2λ and y as a constant)

$\delta L/\delta y = x - 5\lambda$ (because differentiating L with respect to y gives 5λ and x as a constant)

$\delta L/\delta \lambda = 100 - 2x$ (because differentiating L with respect to λ gives 100 and $2x$ as constants).

Set all three of these equations equal to 0 and then it's simple algebra to solve for x (25 units) and y (10 units). Solving for λ gives its value as 5. The variable λ is the marginal utility of a $1 increase in the size of the budget constraint. The problem is extremely simple because we used a simple equation for utility based on two commodities and a simple equation for the budget constraint. For more commodities, the math is more complicated, but the process is equally straightforward.

The same math also enables economists to figure out how to minimize cost. Using the same equation, the agent minimizes the cost of attaining some fixed quantity of utility just by switching the budget constraint equation into the first set of brackets and the utility equation into the second set of brackets:

$L =$ [budget constraint equation] $- \lambda$[utility equation]

Then you differentiate the variables with respect to L, now treated as the minimum cost instead of the maximum utility. Some simple algebra will give its value in units of the prices of the commodities to be purchased.

Ever since Edgeworth, economists have been studding their papers and textbooks with Lagrangians for utility maximization. The Lagrangian is the go-to method whenever the economic theorist describes a problem of maximization subject to scarce resources facing a rational agent. But that's practically Robbins's definition of economics! In fact, students of advanced economic theory are required to learn multivariable calculus primarily to solve Lagrangians for more and more complex constrained optimization problems. From intermediate microeconomics to the most sophisticated exposition of macroeconomic models of growth, you can get a grip on what's going on in the fancy mathematics just by noticing how the Lagrangian is invoked.

Lagrangian equations are ubiquitous in economics because of the way the theory is driven by the proof of the first theorem. The competitive market comes to a general equilibrium, and that general equilibrium is Pareto optimal, because agents solve constrained maximization problems. We will see why in chapter 6 on macroeconomics.

Let's look at some postcards from the edge of advanced economic theory, David Romer's *Advanced Macroeconomics*. We'll see the same Lagrangian used in action in the stratosphere of theory. It's there for the same reason it figured in Edgeworth's work. Economic theory is still committed to the existence of general equilibrium and its Pareto optimality. Romer's book sets out to expound the growth theory for a whole economy. The model starts with utility-maximizing households and profit-maximizing firms that employ the members of the households. Each household has to choose a path of consumption that maximizes lifetime utility, subject to lifetime earnings and initial capital holdings—its budget constraint.

Who Invited Calculus?

The math looks much more complicated than the simple example we worked above, but it's the same, driven by the need to establish a growth path that remains at the Pareto optimal general equilibrium. It's worth getting the hang of these expressions because they show that there is much less to the mathematics of economic theory than there is to, say, Schrodinger's equation in quantum mechanics. All parts of every equation have an intelligible explanation.

Here's the first postcard from Romer's textbook.

$$U = \int_{t=0}^{\infty} e^{-\rho t} \frac{C(t)^{1-\theta}}{1-\theta} \frac{L(t)}{H} dt$$

$$= \int_{t=0}^{\infty} e^{-\rho t} \left[A(0)^{1-\theta} e^{(1-\theta)gt} \frac{c(t)^{1-\theta}}{1-\theta} \right] \frac{L(0)e^{nt}}{H} dt$$

$$= A(0)^{1-\theta} \frac{L(0)}{H} \int_{t=0}^{\infty} e^{-\rho t} e^{(1-\theta)gt} e^{nt} \frac{c(t)^{1-\theta}}{1-\theta} dt$$

$$= B \int_{t=0}^{\infty} e^{-\beta t} \frac{c(t)^{1-\theta}}{1-\theta} dt$$

It looks fearsomely impressive to the outsider. We'll never be able to understand what those equations say! But we will, and when we do, we'll find it hard to take them seriously as descriptions of any economic reality. And we'll appreciate how much economic theory's embrace of calculus is a reflection of its fidelity to the first theorem.

Start with the first line. Despite all the symbols, it's just a utility equation like the one we used above, $U = xy$. In Romer's utility equation U is the total lifetime utility of a household. This utility is to be maximized because the first theorem requires the consumer to decide at time zero on all trades in the future, once and for all. The lifetime utility depends on what each member of the household consumes at each time, $C(t)$, the amount of labor all households sell at each time period, $L(t)$, divided by the number of households, H, to get the amount each sells, $L(t)/H$, which also gives the amount each household can spend. The problem of each household is to maximize total utility over all time, from time zero to infinity. Each household has to compute the ways in which each future period of work and consumption will contribute to the total utility computed at the start. Each household has to discount its lifetime income stream from wages it uses to purchase commodities, and has to decide how much to weight future consumption of commodities now. This is where the number e, a mathematical constant approximately equal to 2.7182 . . . , comes in. Wherever e appears in the equations, it's measuring the present value of some future stream of consumption or earnings. In the above equation it appears twice: as the interest rate, with the exponent ρ, and as the growth rate of labor, with the exponent n. The equation includes the term $1-\theta$ to show that the household weights future consumption as contributing less to lifetime utility than current consumption. The big integral sign after the equal sign, $\int_{t=0}^{\infty} dt$, just means that the household's total utility is summed up from all time periods, zero (0) to infinity (∞).

So what does the first line say?

The household's total utility measured over all time depends on its consumption in each period multiplied by the economy's total income divided by the number of households, since all earn the same amount, with the total future contributions discounted by the compound interest rate. It's a utility equation just like $U = xy$, where x is $C(t)^{1-\theta}/1-\theta$ and y is $L(t)/H$, and there's a discount factor, $e^{-\rho t}$, and an integral sign because it is lifetime utility that is maximized.

Why bother writing down a utility function that no one can compute, and not just because we don't know the numbers? Because the first theorem requires that consumers have this figure at hand when deciding on their commodity consumption, leisure, and labor, and because, as we'll see, macroeconomic models of economic growth require them, too.

What about the next three lines?

Each of the next three equations comes from the first one by plugging in equivalent substitutions or abbreviations to make solving the eventual Lagrangian simpler. You can skip the next four paragraphs in which we make these arithmetical substitutions. They look formidable and they're hard to keep in mind because there are so many symbols, but it's just basic algebra.

The second equation substitutes a big bracket for the individual household member's consumption. The bracket includes the symbol A, for the productive technological knowledge at the start, $A(0)$, which grows with time (look for the e that $A(0)$ is multiplied by). Growing A is multiplied by $c(t)$—consumption per unit of labor (i.e., how much stuff is made in a unit of time), always increasing in its marginal utility (notice the $1-\theta$). The product of $A(0)$ and $c(t)$ is then multiplied by the total labor sold by the household, starting from time 0 and increased by e to the power of nt, where n is the growth rate of the working population. That's how we get capital letter $C(t)$—the total household labor sold—from lowercase $c(t)$ multiplied by the amount of labor one household sells, L divided by H.

Almost there. The third line just moves the constants, the numbers for A, L, and H, to the front of the integral, since constants are never changed by integration. In the fourth equation the symbol B is substituted for the constants as another abbreviation.

Now that we've walked through the equation you can take mathematical economists' word for its form. All that has been done is thinking about how the rational household will decide on packages of labor to sell and leisure in which to enjoy commodities it buys that maximizes its utility over the time period zero to infinity, assuming that utility is always positive and marginally decreasing with quantity, and that the agent's later utility is valued now by some function of e. Now we need the budget equation to combine with the utility equation in the Lagrangian. Then we'll compute the Lagrangian to get the household's lifetime pattern of consumption.

Once you've got a handle on the symbols in the household's utility function, understanding the budget equation it's operating with will be child's play. Recall the simple example above: $0 = 2x + 5y - 100$, the amount to spend and the price of what is consumed.

Romer's consumer still has just a single budget equation, reorganized for mathematical convenience. The expressions are familiar from the ones we just worked through: e to an exponent, this time $R(t)$, which is the result of compounding interest on capital. The household's capital (per unit of labor) at time zero is $k(0)$, and $w(t)$ is the household's wage rate per unit of labor at time t, which increases from time zero at a rate of e to the power of $R(t)$. So the household's budget constraint equation has to set consumption equal to (or less than) the household's starting capital plus the household's lifetime earnings, $w(t)$ discounted into the infinite future via e to the power of $-R(t)$ to give its amount at any time t. In the first equation below, the integral on the left side of the symbol $\leq$ (less than or equal to) is total lifetime consumption. On the right side of the symbol, $k(0)$ is the household's capital at time zero, and the last integral, added to the household's capital, is the household's wage earnings over time zero to infinity. From this equation Romer takes us to the budget constraint he'll use in the Lagrangian to solve the household's lifetime constrained maximization problem. The equation results from simple substitutions you can take his word for.

$$\int_{t=0}^{\infty} e^{-R(t)} c(t) \frac{A(t)L(t)}{H} dt \leq k(0) \frac{A(0)L(0)}{H} + \int_{t=0}^{\infty} e^{-R(t)} w(t) \frac{A(t)L(t)}{H} dt$$

Substituting $A(0)L(0)e^{(n+g)t}$ for $A(t)L(t)$ and dividing both sides by $A(0)L(0)/H$ yields Romer's next equation:

$$\int_{t=0}^{\infty} e^{-R(t)} c(t) e^{(n+g)t} \, dt < k(0) + \int_{t=0}^{\infty} e^{-R(t)} w(t) e^{(n+g)t} \, dt$$

Recall the Lagrangian from our toy example above:

$L =$ [The equation for utility] $- \lambda$[The equation for budget constraint].

This is what comes next in Romer's presentation.

$$L = B \int_{t=0}^{\infty} e^{-\beta(t)} \frac{c(t)^{1-\theta}}{1-\theta} \, dt + \lambda [k(0) + \int_{t=0}^{\infty} e^{-R(t)} e^{(n+g)t} w(t) \, dt$$

$$- \int_{t=0}^{\infty} e^{-R(t)} e^{(n+g)t} c(t) \, dt]$$

The Lagrangian here is equal to the equation for utility from the first postcard plus λ multiplied by the equation for the budget constraint. (It has been rewritten, like the simple example, to make $k(0)$ plus the two integrals equal to zero.) It looks much more complicated, but that's because the Lagrangian solves the household's constrained maximization problem over an infinite number of time periods of labor and consumption.

Of course this equation is never solved for real values of any of the variables or constants, unlike the easy problems in introductory microeconomics textbooks. That's not the point. From here Romer goes on to show that when combined with firms' profit maximization formula, the result is a Pareto optimal general equilibrium for the whole economy over its entire time horizon.

Where Do Equations for Utility Come From?

One mystery raised by any utility equation, however simple or complex, is where it came from. You might suppose that whether it is the simple equation $U = xy$ or any other, we need some factual basis, some evidence, some grounding in data for the relationship between commodities and utility in order to formulate the equation. Before we can consider how utility is maximized, we need to know exactly how it depends on the variables that produce it. If we could measure utility, then we could experiment with varying commodities and recording utilities to formulate the required equations. But there is no such thing as utility. It went the way of caloric and impetus. The absence of utility should make mysterious the equations for utility that are plugged into the Lagrangian.

Some of the simple equations in the textbooks for how utility depends on commodities are in fact unmysterious. But that's not because they result from empirical investigation. They are the results of thought experiments undertaken by economists and their readers. For example, if two goods are substitutes, like butter and margarine, introspection suggests that as you consume more of one, you consume less of the other. Both give you some satisfaction, but the more you have of one, the less you need of the other. We could express this commonsense truth mathematically and produce an equation for how quantities of utility depend on the two commodities, if there were such a thing as utility. The utility equation would make U a function of the quantity of butter plus the quantity of margarine: $U = f(b+m)$. When goods are complements, like hot dogs and mustard packets, introspection suggests you consume them together, and if you run out of one, you don't consume more of the other. So the equation for utility derived from complements must be something like $U = f$(lesser amount of the two goods if you have some of both), $U = f$(Min [number of hotdogs, number of mustard packets]). If your consumption of one good were independent of your consumption of another, the equation for the utility derived from both, if there were such a thing as utility, might take the shape of a Cobb-Douglas function: $U = A(x^a y^b)$.[1] Our simple Lagrangian example was a Cobb-Douglas utility function in which A, a, and b are all equal to 1, because in this equation increases or decreases in either good increase or decrease total utility independent of the other good. (Romer's equation for a household's lifetime utility was a Cobb-Douglas function, too.) Still another utility equation, the quasilinear utility function, is created from a thought experiment in which your preferences of one commodity over another, say leisure time, t, over food, c, remains constant, no matter how much of the latter one consumes: $U(t, c) = v(t) + c$, where $v(t)$ increases because you always prefer more leisure to more food. These simple utility equations, often found in the problems at the back of textbook chapters on constrained maximization, are all formulated by thinking about how people consume combinations of commodities that they know something (often a lot) about. Unlike the equations that relate variables in the ideal gas law or its successors, these equations don't emerge from empirical data. So we can't cash in the f in $U = f(x, y, \ldots)$ for a particular mathematical relation between its variables.

There really is no such thing as utility to be maximized, no utility that comes in marginal quantities, no utility that can be put into equations to describe how its quantity varies with consumption inputs. But proving general equilibrium needs equations that can be treated by differential calculus to solve constrained maximization problems.

Utility as a Mathematical Convenience

Economic theorists insist that we can make use of the mathematics without taking the equations seriously as describing any facts about utilities. The equations are just mathematical conveniences economics uses to express and systematize ordinal preferences, which, after all, are the only things there are for consumers to care about.

Economists recognize that there can't really be equations relating quantities of utility to consumption bundles; there can only be rankings of bundles. They square the continued use of quantity of utility equations even when there are no quantities of utility by invoking a mathematical device for saying that the *numbers in the utility equations don't matter*. Given a preference order there will be an infinite variety of different equations for utility that all describe the same ordering. Treat them all as equivalent; it doesn't matter which one you use. None of them is more correct than any other. None is less correct than any other. If you use them the right way, you won't find yourself committed to measurable utility even though your equations make it look like you are.

Here's a simple example. Take any utility equation

$U = U(x,y,z, \ldots)$.

Multiply the $U(x,y,z, \ldots)$ by any constant A and add any constant B:

$U = A(U(x,y,z, \ldots)) + B$.

In math-speak this is a *positive monotonic transformation* of the original equation. If you graph the first equation, then apply constants like A and B to it, the resulting graph will connect new points in the same order of increase, though with different distances between the new points on the graph. This is all economic theorists need in order to insist they're interested only in preference rankings while still helping themselves to differential calculus to describe consumers' choice behavior.

The economic equivalent of the positive monotonic transformation of a utility equation is just a way of saying no psychological fact corresponds to any one utility equation. Utility equations don't describe actual quantities behind preferences. The only reason we use them is to employ calculus to solve constrained maximization problems that theory tells us economic agents solve. But we can't take the equations seriously, the way we can in a domain where equations express something more than stylized facts from which other, non-stylized facts will be derived.

In most domains of science, using differential calculus to calculate the real value of a causal variable is the gold standard for confirming a theory. Among physicists, "shut up and calculate" is something of a watchword, and Nobel Prizes are given for calculations that pan out. The billiard ball model enabled the calculations of the values of pressure, temperature, and volume that experiments established empirically. The general theory of relativity enabled Einstein to calculate the orbit of Mercury, which had been established by observation. In natural sciences, math enables us to derive numerical facts, not just stylized ones. The derivation of a number is what establishes the theory's explanation as at least on the right track.

But economic theory doesn't use calculus this way. It can't, because deriving specific numbers is impossible unless one of the infinite number of positive monotonic transformations is the correct equation, the one to which mathematical tools can be applied to calculate a value to test. But there is no number—no quantity of utility—to calculate. So what do all the equivalent utility equations do? Any one of them lets us derive the stylized facts about downward-sloping demand and upward-sloping supply that the theory needs. And because the theory is limited to explaining these stylized facts, it doesn't have to engage in the calculations that other sciences use calculus to attain.

Once economic theory gets past the most basic utility equations—for complements and substitutes, quasilinear relations among commodities—it helps itself to whatever utility function will enable it to derive the particular stylized fact that is to be explained. Often the utility function chosen has a Cobb-Douglas form, $U(x,y) = Ax^b y^c$, as this equation, or any positive monotonic transformation of it, will make the derivation of a general equilibrium outcome possible, and even sometimes mathematically easy.

The Calculus of Stylized Facts

As we've seen, all economic theory needed to prove the first theorem is that consumers' indifference curves are convex from the origin. When indifference curves were thought of as resulting from utility, the requirement for utility equations was only that they imply convex indifference curves. When economics stopped addressing the question of how convex indifference curves arise, it didn't need utility equations, or at least it didn't need to treat them as descriptions of reality, whether economic or psychological.

But economic theory still needs utility equations to do its work. It formulates sets of these equations—positive monotonically equivalent equations. It plugs an arbitrarily chosen one of them into a Lagrangian in order to derive the stylized fact that economists seek to explain. There is, in these explanations of stylized facts, an implicit understanding that even though every individual has only ordinal preferences, when we aggregate large numbers of individuals' preference choices together, we get increasingly smooth curves to which we can harmlessly apply differential calculus to derive stylized facts.

Typically, an economic theorist will work backward from a stylized fact to an equation for the consumer's constrained utility maximization problem, from which the stylized fact can in turn be derived mathematically. But there is no question of empirically testing any single equation for a quantity of utility produced by consumption, either via observation of actual consumer choices or in experiments on real people.

Economic theorists repudiate utility as unreal while embracing the word, the symbol, in equations. But this raises all the same problems about idealized models we explored in chapter 2. The usefulness of idealization can't be a matter of enabling us to predict actual choices. The usefulness of utility can't be that it explains actual choices, since we already know there is no such thing as utility. Usefulness could be a matter of mathematical tractability. That is, adopting the fiction of utility enables us to use the mathematics of differential calculus to derive the stylized facts economists recognize. Without Lagrangian methods and the required equations of the form $U = f(x,y,z, \ldots)$, economic theory cannot derive some of its most important results. This includes the grounding of the whole economy's operation on the RCT choices of all the individuals in the economy, as we saw in chapter 3 and this chapter.

Without a substance that comes in continuous quantities and can be subjected to differential calculus by its theory, there is not much of a predictive or explanatory role in economics for the formidable mathematics it (alone of the social sciences) proudly shares with natural science. This math is what underwrites its status as science, among outsiders to economics, by comparison to the other social disciplines. More than one economist (e.g., Paul Krugman) has diagnosed the problems of economic theory as due to it substituting the beauty of mathematical theorizing for messy truths about the real economy. Here we have a good example of how a concept, utility, sticks around in economics long after economists have sworn it off, just because of the mathematics it allows the theory to flaunt.

5 Money—Who Needs It? Not Economic Theory!

> Hamlet without the prince: a performance or event taking place without the principal actor. The phrase comes from an account given in the *Morning Post* of September 1775. The member of a theatrical company who was to play Hamlet in a production of Shakespeare's play ran off with an innkeeper's daughter before the performance; when the play was announced to the audience, they were told "the part of Hamlet [was] to be left out, for that night."
> —*Farlex Dictionary of Idioms*, s.v. "Hamlet without the prince"

Nothing is more obviously a part of the economy than money. But there is no room for money in microeconomic theory. The development of economic theory from rational choice theory onward does not require it. And when it is introduced, it plays no real role in the working of the economy. This is an unavoidable consequence of the role of RCT in microeconomics. As a result, when we seek to understand what money does, or what makes something qualify as money, economics provides no answers, and furthermore proudly insists it need not do so. These are not questions for economic theory. As we'll see in this chapter and the next, the exclusion of money from RCT as a real factor in the economy twists economic theory into knots.

Real, Relative, and Money Prices

It's important to understand why RCT leaves no room for money in the exchange economy. The rational agent's preferences are for consumption goods—things and services needed to live or wanted to enjoy life. The agent trades with others until the market reaches a Pareto optimal distribution. But the *until* in the previous sentence is misleading, since the theory presupposes either that all trades are instantaneous or that none are made until all bidding has been completed. More importantly, the only prices relevant to trades between rational agents are what economists call *real prices*: the ratios of amounts of goods that traders will accept in exchanges—say, one apple for two bananas. It's a potential matter of confusion that these real prices—these ratios of goods to goods—are sometimes called *relative prices* (in other disciplines, relative values are contrasted with real ones). But in economic theory, real or relative prices are the only ones that matter to rational agents balancing their consumption of different goods and deciding how much of each to acquire. In the perfectly competitive economy, every RCT rational agent has complete information and knows all the real prices—all the goods-for-goods trading ratios that prevail at the general equilibrium.

Of course, rational agents trading on a market in something like real time will need money. It's a tool for solving the evident problem of barter (the moneyless exchange of goods). This is the problem of the double coincidence of wants, or rather the absence of such a coincidence. Bartering often requires an exchange of goods that are hard to divide into small-enough units. Trading an apple for one and a half bananas leaves the banana supplier with half a banana that may be hard to get rid of in another exchange before it goes bad.

The economy in which the first theorem proves the existence of a market-clearing Pareto optimal general equilibrium doesn't have this problem. All commodities in it are infinitely divisible. But substitute for the assumption of infinite divisibility of commodities the more realistic assumption that they are lumpy and difficult or even impossible to divide. Then, if they are smart enough, rational agents will invent money to solve their barter problem. But they will have to get enough other rational agents to agree to a binding commitment to exchange all other commodities for the one commodity everyone agrees to use as money—not a trivial problem. But assume it can be solved.

To agree on what would work as a solution to this problem, rational agents need to find a commodity with these features:

1. Almost everyone desires it for immediate consumption, and knows that almost everyone else does, too.
2. It's durable and doesn't wear out when handled.
3. It's easy to carry around in sufficient quantities for trade.

4. It's relatively easy to divide precisely.
5. Its total supply is limited and quantities of it are hard to come by.

If a sufficient number of agents engaged in trade across markets for different goods agree on something satisfying these conditions, the traders can solve the problem of the double coincidence of wants. If all, most, or even just many people like collecting and admiring bits of shiny metal, then these bits will fill the bill. The price of anything, measured as a ratio of its amount to an amount of shiny bits of metal, is just another relative price. As we'll see, once used to solve the barter problem, there is no way for money to change anything in the economy; there is no way for it to cause anything to happen.

Evolutionary Just-So Stories About Money?

People are not perfectly rational or even very farsighted. They didn't intentionally design coinage to address the double coincidence of wants. They might have surmised that almost everyone would enjoy collecting and contemplating bits of shiny metal. But no one convinced everyone else that bits of shiny metal satisfied the other four requirements. There had to have been a process of cultural selection that converged on bits of metal as money. It wasn't the only candidate solution. Cattle and sheep, blades and axes, and shells and slaves served as money in various cultures at various times. Trade for bits of shiny metal emerged independently in several regions of the world and outcompeted other possibilities because bits of shiny metal were a better solution to the barter problem. Societies where the institution of money in some form didn't emerge couldn't trade easily and so didn't thrive, prosper, or survive. Coinage won out in a process that repeatedly and independently, though gradually, selected for a locally optimal solution to the double coincidence problem—without anyone intending it.

How money evolved from bits of shiny metal to pieces of paper and then to bits in a computer's memory bank is another matter. The real history of the emergence of money may involve a very different scenario from the just-so story one might tell about how Darwinian cultural selection converged on an efficient solution to the problem of the double coincidence of wants. One alternative historical account doesn't start with money as a solution to the barter problem, and it has a nice explanation of how paper money arose. The story begins with the hydraulic despotisms that arose after the development of agriculture ten thousand years ago. Mesopotamian empires extracted wealth from the economy by taxing agricultural production. When the government's tax authorities accepted only wheat (or only some other product) as payment, it quickly became a commodity everyone accepted in trade, due to their need to pay the tax. This would have immediately solved the common knowledge problem—everyone would have known about the tax. For this rather inconvenient unit of tax payment, amounts of some other commodity could be substituted by the tax authority or ordinary people if they were confident in its acceptability to those who had to pay in the original coin of the realm—say, wheat sheaves. Carrying around wheat sheaves is inconvenient. If the authorities began to accept written promises to bring wheat sheaves in as tax payments, the written promises would be accepted by others who had to pay tax in exchange with each other.

This story substitutes the conscious design on the part of rulers to exercise taxing power for the unintended creation of money in barter. But then the process of Darwinian cultural selection takes over, using wheat sheaves, or promises to provide them, to solve the persistent problem of the double coincidence of wants.

Probably money has more than one origin—as a solution to the barter problem and as the only thing the Mesopotamian empires would accept as payment of taxes. Cultural anthropologists have uncovered a trove of different things used for money in early recorded history (and even earlier), before the documented emergence of coins in Lydia (today's Turkey) around 600 BCE and in China a few hundred years later (where paper money also emerged almost at the same time).

Economic theory doesn't need to know where money came from. That's history, not science. Economics might not really even need a definition of money. After all, mathematics and physics have no need for definitions of the concepts indispensable to their expression. Number theory doesn't define *number*, and apparently the progress of mathematics over the last two millennia has not been impeded by this oversight. Similarly, *mass* is undefined in physics. In economics *money* is not a basic term employed to define other concepts; rather, it describes a convenience that rational agents could agree to bring into existence as the solution to a problem. More importantly, if money is *causally inert*—has no effects in the economy—then it doesn't matter to economic theory how it originated or what qualifies as money. If what counts as money and the amount of money sloshing around the economy don't matter to people's economic choices or their willingness to trade commodities, then economics doesn't have to take sides on questions of history—if, when, and how it is created or destroyed. On the other hand, if what counts as money and how much of it there is does affect the economy, then money has a real causal role that economics needs to identify and quantify. In that scenario, figuring out what money is, how it is created, and how it is destroyed will be a job for economic theory.

Fortunately for the theory, this didn't turn out to be the case. Money is *causally inert* in modern economic theory. It's enough for economic theory that money is something that really smart people can minimize the use of, or even do without. As we'll see, in the standard model, rational agents never carry around a penny more money than they need to do their shopping. Money has no real, indispensable, or independent role causing anything microeconomics needs to explain. Microeconomics has neither room nor need for it.

The Store of Value?

Once we have money in an economy as the (almost) universally accepted solution to the problem of the double coincidence of wants, another role for it becomes available. It's crucially important, though still a convention, a construct, introduced by rational agents in order to accomplish something "real." This is the function of money traditionally described in economics as *store of value*. It's important to see that this function of money still doesn't make it a force in the economy; it's more of a bookkeeping convenience.

Most definitions of the store of value are unhelpfully tautological; they just say that money maintains its value over time. But *value* is the term that needs explication. What is value in a domain where there are only preferences? Nothing in the economy is intrinsically valuable. What makes something function as a store of value that endures over time is its status as *a claim on future consumption of anything and everything*. What economists are describing is something that by sufficiently widespread and ongoing common agreement will be accepted in the future in exchange for immediately consumable goods.

Rational agents can, at least sometimes, produce more than they immediately consume. Why should they put in the extra effort to do this, especially if what they produce will spoil before they can consume it, like many agricultural products? They will do so only if they can trade the spoilable stuff now for things to consume later. On the other hand, some rational agents produce commodities that require more time before they can be exchanged for goods for immediate consumption. These agents need to feed, shelter, and clothe themselves while producing such commodities that don't spoil or don't spoil quickly. Think of the production of tools—a plow may eventually help to increase crop yields, but its maker has to eat in the meantime. Many commodities that require lengthy production processes will be *capital goods*—things that enable the production of consumption goods. Producers of these items will eventually be able to trade their goods with those who have produced surplus consumption. But meanwhile they have to eat.

Now, why work to produce commodities beyond your immediate needs? The obvious economic motive is self-interest: the only reason you would postpone consumption of some good in your *endowment*—what you already have—is if you get more of it back in the future than you give up now. You need to get more back later to give up what you have now for several reasons. First is the risk that if you give up something now, it won't be given back to you later. Second, your needs, tastes, or preference might change such that the amount you give up now won't bring as much satisfaction in the future. As behavioral economists have shown, there is a solid evolutionary explanation for why we prefer immediate consumption over postponed consumption (recall hyperbolic discounting from chapter 2). In the environment of early human adaptation, among hunter-gatherers, resources were inconsistently available and you could never be confident about the timing of your next meal. So there was Darwinian natural selection to prefer immediate consumption over deferred, uncertain consumption.

In economics there is a name for the prices paid by those who need to acquire immediate consumption goods produced by someone else while they work on goods that take a long time: the *interest rate*. Like all real prices, interest rates reflect ratios of consumable goods. The longer you have to postpone consumption, the higher the price you'll want to do so. The longer it takes to produce a good, the greater the demand for someone else's postponed consumption good.

As noted above, the commodities it takes a long time to make are mostly capital goods, used to produce immediate consumption goods and to improve the productivity and efficiency of immediate consumption goods. In other words, postponed consumption makes economic growth possible.

How much consumption should be postponed in order to provide for other immediate consumption? Like all scarce commodities with multiple uses (recall the definition of economics as the study of this question), the answer is given by the *price mechanism*. The price of postponed consumption of some commodity is the interest rate for loans of that commodity on the market: How much more of that commodity do you need to give back on a certain date in the future in exchange for some quantity of that commodity now? We can use the Edgeworth box from chapter 3 to show that, among rational agents, there will be an exchange market for any particular postponable consumption good that results in exchange at a real, Pareto optimal price. That price will be the real interest rate for that good in that market. It will be a futures market, in which what is bought and sold is the delivery of a commodity at some future date.

The problem of the double coincidence of wants is even more likely to arise in this futures market for postponed consumption than in exchanges of immediate consumption goods. This is where money comes in as an *intertemporal* (from an earlier time to a later time) *solution* to the problem. If a pound, euro, dollar, or yen is accepted today as a permanent claim on a certain amount of any consumption good at any time in the future, then people will want money so that they can trade commodities now for later consumption of commodities.

Once money is widely enough believed to be a permanent claim on the future consumption of any immediate consumption good, rational agents will invent something new, built on the invention of money. They will build what economists call a *bond*: a promise to pay more in the future for some amount of money (or other consumption good) now. A bond is just a tool, a convenience among rational agents, like money. Rational agents are interested only in relative or real prices—the ratios of exchange between all pairs of commodities and ratios of amounts of consumption goods for immediate and postponed use (the real interest rate)—not monetary prices—the ratios of exchange between bits of shiny metal (if that's what money is) and all other immediate consumption goods. But to solve a now-versus-future double coincidence problem—immediate consumption of goods against postponed consumption—they resort to bonds that promise interest payments in units of money.

But money still doesn't have a real causal role in an economy comprising rational agents. It's still just a convenient bookkeeping device. What's more, it's obvious to the economist that no rational agent would ever carry money around unless it was absolutely, unavoidably necessary. Why?

When rational agents can trade postponable consumer goods beyond their immediate needs for bonds that pay interest, they will do so. To fail to do so is to forgo larger quantities of future consumption, which violates one of the assumptions of rational choice theory: that more is always preferred to less. Keeping money in your pocket when you could use it to buy bonds is irrational. Rational agents won't keep a penny more than they need in their pockets.

How much will that be? To answer that question, economists add a few new assumptions to the model: First, that buying and selling bonds costs something, what financial institutions call a brokerage fee. This is what economists call a *transaction cost*: what traders pay to whoever set up the market to be permitted to exchange one good for another. Second, that the more you exchange, the lower the price, so transaction costs go down as the amount of money agents want to trade bonds for increases. The rational agent calculates the cost of holding any amount of money as the sum of the transaction cost of trading a bond for money plus the interest forgone during the time the agent possesses the money. The money withdrawn is used for purchases of immediate consumption goods. If the consumer withdraws money from a bank—that is, exchanges postponed consumption that has been earning interest for money—but does it only once a year for all purchases in the year to come, the amount of forgone interest will be great, though the transaction cost of such a large withdrawal will be low. But the total cost of an annual withdrawal may be much higher than withdrawing money twice a year or every month. The graph of transaction cost plus interest forgone reveals that in many cases, more withdrawals of smaller amounts may be preferred to fewer withdrawals of larger amounts (see figure 5.1).

In order to determine how often to withdraw (i.e., to sell interest-bearing bonds for money), the rational agent calculates an optimum cash balance—the quantity of money to hold that minimizes cost in terms of future consumption (see figure 5.2).

The transaction cost curve slopes downward: the more you keep in cash, the less you spend in transaction costs. The *opportunity cost*—the interest not received when holding money—slopes upward: the more you keep in bonds, the more interest you earn. The two curves combine to produce a curve of the combined cost of holding cash. The rational agent minimizes this cost to hold the optimal cash balance.

Money—Who Needs It? Not Economic Theory!   81

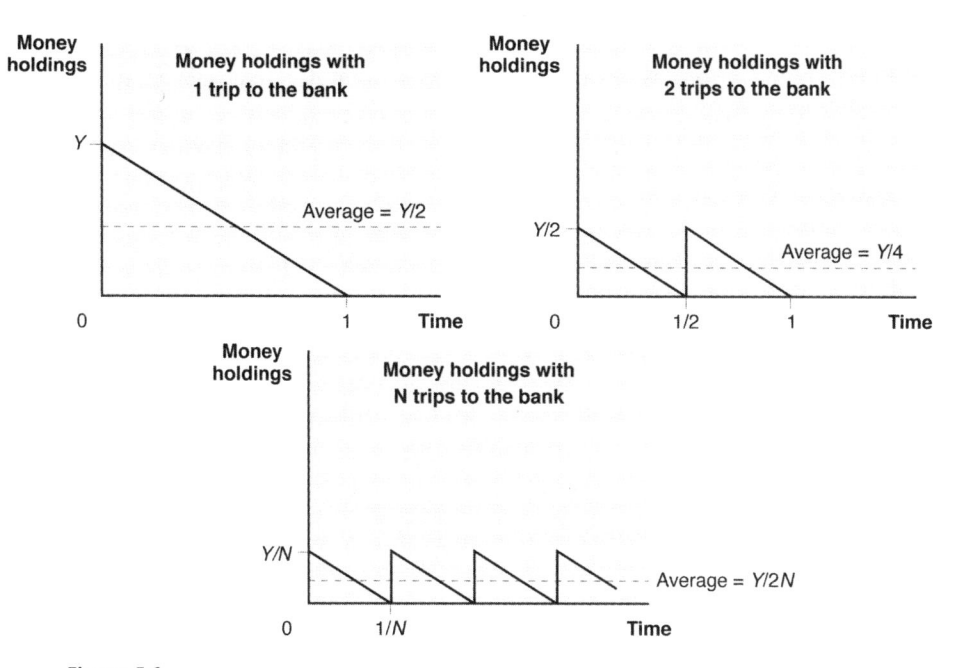

Figure 5.1
Money holdings as a function of frequency of exchange of bonds for cash.

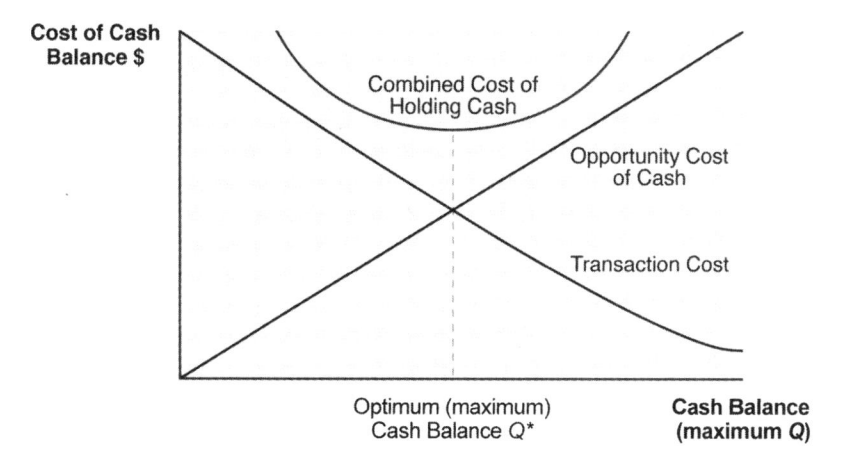

Figure 5.2
The equilibrium optimal level of cash held by the rational consumer.

Once bonds are invented and agents can lend and borrow with interest, the store of value role for money disappears. Money is reduced to a device for solving the double coincidence of wants problem between current and future consumption. Thus, among rational agents, money is just an accounting tool. It doesn't have a real role in the economy. Microeconomic theory forbids it on pain of surrendering the assumption that people are rational.

There is a little payoff to this theory in the neat explanation of another stylized fact. For a long time there have been data showing an inverse correlation between an economy's interest rates and the demand for money—people withdrawing cash from bank accounts. The downward slope of the transaction cost and the shape of the combined cost curves in figure 5.2 are stylized facts that can be derived from the model. But the numerical value of the slope of the curves connecting the transaction cost and opportunity cost variables, even if we could calculate them, is something the model can't explain.

The RCT-driven account of the role of money can retrospectively explain other stylized facts as well. The advent of ATMs and credit and debit cards significantly reduced the transactions costs of converting interest-bearing bonds (in the form of dividend-paying shares, savings accounts, money market accounts, and so on) to cash. In fact, it has eliminated the role of cash as an intermediary in exchange altogether. People have begun carrying much less of it; in some Scandinavian countries no one carries cash at all. Economists extract a stylized fact from recent data about the spread of credit cards and the installation of ATMs. Then they derive it from their models. As the height and slope of the transaction cost line declines, the optimal cash balance point on the graph shifts to the left, toward zero cash balances.

It's too much to expect of economic theory that it predict the emergence of new technologies and the quantitative value of their impact on the economy. Technological change is an *exogenous variable*—a change from "outside" the economy that the market reacts to in unpredictable ways. The most we can expect is the derivation of stylized facts extracted from data after the fact. What has happened in Sweden is an example.

Economists might hope that the technological changes that have reduced people's interest in holding any money—bills, banknotes, coins—at all will continue until money becomes as unimportant in the real economy as it is in economic theory. There will still be bonds—claims on future consumption with interest—so long as there are differences in production time among commodities, especially for capital goods whose utilization increases the production of consumption goods. In the real economy there will be many different kinds of claims on future consumption—stocks, bonds, loans, mortgages, money market deposits, interest-bearing checking accounts. But all of them will work like the simple interest-bearing bonds of the monetary model, and trading them will carry different transactions costs, making some much more *liquid*—more quickly tradable—than others. Rational individuals will distribute their claims on postponable consumption differently among them, depending on the amount of immediate consumption they choose to engage in. The economist will have a ready model to account for this. It will be the same one used to explain the optimal cash balance, now applied to derive a similar graph for the optimal holding of any exchangeable interest-bearing claim on future consumption. The amount of each of these claims an agent holds will depend on the transaction costs of trading them and the higher interest of less liquid bonds that must be forgone to hold more liquid ones. Whatever these amounts are, they will play real roles in the economy, being caused by and causing differences in rational agents' preferences about their willingness to consume immediately or to postpone consumption.

The Neutrality of Money versus the Money Illusion

Economists express the exclusion of money as a force in the real economy that microeconomic theory requires by insisting on the *neutrality of money*—the assertion that the amount of money in the economy and changes in that amount have no causal impact on the economy. The flip side of money's neutrality is that people do not suffer from a *money illusion*: they don't change their behavior as a consequence of the amount of money at their disposal. Rational choice theory insists on the neutrality of money. But economists know that people do change their behavior depending on how much money is in their pockets, paychecks, and bank accounts. So they need to explain this stylized fact in a way that is consistent with the RCT-driven theory that forbids people from suffering a money illusion. How do they do it?

The neutrality of money, its inertness as a factor in people's consumption and production decisions, follows directly from RCT. Since rational agents make choices about trade solely on the basis of relative prices, the actual number of dollars, euros, pounds, or renminbi in people's possession will not influence their purchases. If the amount of money in an economy doubles overnight and everyone knows it, and if the only role of money is to solve the barter problem, everyone will just pay twice as much as they paid before, leaving relative prices exactly where they were. Whence the neutrality of money. The same thing happens if the amount of money everyone has is reduced by half: there is no impact on real prices even as everyone pays half as much money for the same goods. That's why economists devoted to the explanatory power of RCT have a strong motive to seek evidence that the real economy doesn't respond to changes in the amount of money.

Even rational agents are not likely to follow day-to-day changes in the quantity of dollar bills or euro notes in the real economy. Still less are they going to know what's in the wallets of everyone around them. Without that information about everyone else, the rational agent may react to a doubling of the amount of money they have as an opportunity to buy more, increasing demand and thus raising prices. If they discover that they have half as much money as they thought, the opposite will happen. If uniform changes in the amount of money are not universally known they can have quite an impact on the economy, even if everyone is acting rationally. Whence a temporary money illusion lasting until all rational agents finish bidding up (or down) the prices of everything. Once they do so, all relative prices will be back where they were before everyone's cash balances doubled or halved. A short-term money illusion gives way to long-term neutrality of money.

Long term and short term are important, recurring concepts in economic theory. It might seem natural to ask how long each sort of period is, maybe even ask how many short terms are in one long term. But that's not how the two time periods are defined in economic theory. Unlike the physicist, who calibrates a clock first and then uses its units to test a theory, the economist works backward from theory to identify the units of time. The long term is the amount of time during which the economy is at or near its Pareto optimal market-clearing equilibrium. Since at this equilibrium no trading occurs, nobody is holding a cash balance. There is no role for money. The short term is any period in the long term during which the economy is departing temporarily from equilibrium: prices are changing. Whether the price changes are caused by real shifts in demand and supply or just changes in the amount of money sloshing around in the economy cannot be immediately determined by rational agents. When they get things wrong, they will suffer from a short-term money illusion.

To calibrate the short term and the long term, economists first establish the existence of a general equilibrium and then measure time intervals on that basis. The money illusion obtains during the short terms when the economy is not at or near general equilibrium. During these periods, people mistake higher (or lower) money prices for higher (or lower) real prices and change their behavior accordingly. As they learn the error of their ways, their behavior gradually returns to its long-term levels of general equilibrium.

The way in which economic theory deals with the neutrality versus illusion phenomenon reveals again how difficult it is for theorists to get beyond stylized facts in their explanations and how daunting this makes offering predictions for the real economy. Measuring correlations in calendar time between changes in the money supply and levels of economic activity is not difficult. But converting short-term and long-term periods into calendar units—months, quarters, years—requires first that we can tell whether and when the economy is in, near, or headed in the direction of general equilibrium. That's the basis on which the distinction between the long and short term is drawn. It's difficult to pin down when the economy has arrived at or how far away it is from this equilibrium, so it will be difficult to use real-time data to get beyond "explaining" the stylized facts whose existence is derived from and assured by the assumptions of microeconomic theory.

Once, Long Ago, Money Was Real

Long before the emergence of RCT-driven economic theory there was a model of the economy that gave money a real role. It was used to explain the phenomena of inflation and deflation, changes in money prices, and sometimes the growth of the economy.

The *quantity theory of money* is a simple equation that goes back hundreds of years to long begore the emergence of microeconomic theory in the nineteenth century.

$MV = PQ$

expresses a relationship that was formulated by no less a philosopher than David Hume, echoing an insight that supposedly went back through John Locke to Nicolaus Copernicus. It states that the total quantity of money in an economy multiplied by the frequency with which a unit of it changes hands in trade (its velocity) is equal to the average money price of all the goods and services traded in the economy multiplied by how much is sold (the quantity). The equation $MV = PQ$ made sense among thinkers about the economy at the time. If people use money mainly to buy things, and if they buy everything produced at some price or other, then each unit of money will have to change hands enough times to cover all the trades.

The quantity theory of money is the first appearance of macroeconomic theory, hundreds of years before Adam Smith began to inspire microeconomics. It expresses a relationship between four macro variables, or four features of the economy as a whole: the total quantity of money in the economy, the average price level for all goods and services, the mean rate of movement of all banknotes or bits of shiny metal from one pocket to another, and the total quantity of goods bought and sold.

The way it works today, $MV = PQ$ is an equation that in important respects resembles $PV = kT$, the ideal gas law we explored in chapter 2. Drawing out some of the similarities will help us understand the differences between models in economics and in physical science.

Like $PV = kT$, the quantity theory equation presents an instantaneous, snapshot view of the relationship between its variables. Consider pressure and temperature, P and T. Both are the result of the velocity of molecules hitting something: the sides of the container, for pressure, and other molecules, for temperature. Changing the pressure of the gas changes its temperature in the same instant. To change pressure, make the molecules move faster. That changes temperature at the same time, and vice versa. We know why: both are matters of the motion of the molecules. The billiard ball model of a gas causally explains the instantaneous correlations that $PV = kT$ records.

Something similar happens with $MV = PQ$. The equation tells us these four variables move in lockstep. But they don't adjust second by second, or even month by month. They have to march together over the long term. We need a model of the economic mechanism that keeps them in lockstep. And just as chemistry sought a causal mechanism to explain the ideal gas law and found it in the atoms of the billiard ball model, economics wants an explanation of the quantity theory in some underlying processes involving economic agents.

From early on, the quantity theory was used to explain persistent inflation over several centuries that followed the European conquest of the Western Hemisphere, especially after the influx of gold and silver from Spanish colonies. This long-term, persistent, sometimes rapid increase in the money supply, M, was accompanied by increases in the average price of almost everything, P, even as the total quantity, Q, of things produced grew slowly. The philosophers and physicists (there weren't any economists yet) who articulated the quantity theory saw the chain of causation for inflation as starting with increases in M. It was natural to consider V—the velocity or rate of circulation of the coins—as a constant that reflected people's periodic engagement in trade to sustain themselves. So if Q remained the same, it was only prices, P, that could rise to keep the $MV=PQ$ identity stable.

The quantity theory equation reflected a real causal role for money in the economy. Those who expounded it treated the equation as part of an explanation of how earlier events, such as influxes of gold and silver—changes in the money supply—cause later events, such as increasing prices. These, in turn, will stimulate production and eventually overproduction that will bring prices down again. Meanwhile, money supply changes encourage people to use their money more quickly to avoid inflationary losses in its value before they can get rid of it, thus increasing the velocity of money. Hume argued that people suffer from the money illusion. Sellers treat an increase in the amount of money that comes into their possession as a signal of increased demand, so they raise prices and increase output, thus producing economic growth. Increases in the money supply and consequent inflation have other effects, benefiting debtors at the expense of creditors, which encourages borrowing, which expands production and trade further until they go too far and bring about a deflationary bust.

Nineteenth-century economic theory put a stop to the quantity theory as a macroeconomic explanation of the business cycle, the boom and bust of real economic processes. It had to. The rational choice theory that guaranteed the Pareto optimality of the general equilibrium deprives money of any causal role in the economy. All money does is solve the double coincidence of wants problem, so no interpretation of $MV=PQ$ that gave the money supply real power to change total production, Q, was going to be allowed. If money is neutral, if people don't suffer from a money illusion, then changes in the money supply won't change total economic output. The same goes for the money-price level. Rational agents who care only about relative prices will be unaffected by changes in average money prices. Doubling the money supply will double the money-price level and halve the velocity of money transactions by rational agents, but it won't affect Q—what the economy actually makes and trades. And that is what is important, not the mere bookkeeping of money supply, price level, and velocity.

Of course the quantity theory equation, $MV=PQ$, remains in force among rational agents. It's just causally inert, irrelevant in the explanation of real economic changes. As we've seen, rational agents won't hold a penny more or less than they need to do their shopping, buying everything at its equilibrium price. (This is the price at which there is no excess demand or excess supply; nothing at that price is undersupplied or left unbought.) Rational agents adjust their individual shopping schedules and the velocity of their transactions so that the money supply, however much it is, is just enough to make all necessary purchases. Money supply will change money prices—more money, higher money prices; less money, lower money prices. But it won't change real, relative prices.

Having sidelined the quantity theory of money, economics deprived itself of one resource that had been used to explain the boom and bust of the business cycle. In the face of repeated financial panics and ensuing depressions in the nineteenth and early twentieth century—in the United States in 1819, 1837, 1873, 1893, 1907—economic theory had no tools to explain their severity or duration. In fact, economists had to rule out economic explanations for their occurrence altogether. The onset of these panics could be shrugged off as the result of noneconomic exogenous factors, like the weather or war or new inventions that economic theory had no scientific obligation to account for. But economists were chagrined by having to maintain a mystified silence about how deep the subsequent depressions were, how long they would last, and what processes might bring them to an end. For example, they couldn't explain the duration of economic slumps by asserting that deflation reduced demand and thereby also reduced total output. That would be giving the money supply a causal role in the economy via the quantity theory.

In the face of repeated panics and the slumps that resulted, the most reassurance the theory could give was that in the long term the economy would return to a Pareto optimal, market-clearing general equilibrium. Of course market clearing includes the labor market. Once wage rates fall enough to clear the excess supply of labor—people willing to sell it but no buyers for it—there will be jobs for all who want them. End of depression.

How long is the long term? How long is the short term? Recall the role of theory in identifying these lengths of time. The long term is time during which the economy remains at or near a Pareto optimal general equilibrium. The short term is a smaller part of this long term during which the economy isn't at equilibrium but is moving toward it. Notice this way of defining the long and short terms makes it hard to argue with the statement that in the long term the economy is at or near a Pareto optimal general equilibrium.

Keynes Tries to Make Money Real . . . and Fails

This is where John Maynard Keynes comes into the story of economic theory. Keynes was concerned that classical economics could not explain the duration, or provide tools to mitigate the harms, of the Great Depression of the 1930s. Even before that he had been skeptical of classical economic theory's commitment to the sort of Pareto optimal, market-clearing equilibrium driven by arguments based on Edgeworth boxes. More importantly, he had argued in *A Tract on Monetary Reform* (1924) that V and P can vary over time along with M and at rates different from those that keep all three moving in synchrony as required by RCT.

Keynes accepted the quantity theory as a long-run truth. But its long-run truth was qualified by one of his most famous observations:

> An arbitrary doubling of [the money supply] . . . must have the effect of raising [the price level] to double what it would have been otherwise. The Quantity Theory is often stated in this, or a similar, form.
>
> Now "in the long run" this is probably true. . . . But this *long run* is a misleading guide to current affairs. *In the long run* we are all dead. Economists set themselves too easy, too useless a task if in tempestuous seasons they can only tell us that when the storm is long past the ocean is flat again.[1]

Finally, in *The General Theory of Employment, Interest and Money*, Keynes introduced a causal role for money beyond being a convenience in trade. He also insisted on a causal role for money prices, the P on the right side of the equation, and on people's willingness to part with money, affecting its velocity, V. To do so Keynes had to make a radical break with the foundations of RCT—a break that has not been fully appreciated.

Besides its role in exchange, Keynes argued that money had a "speculative" role in the economy and a "precautionary" role as well. People are willing to hold money above and beyond their use of it to buy stuff. These two roles of money reflect a fundamental fact about the reality that real economic agents face—radical, probabilistically unmeasurable *uncertainty*. In Keynes's words:

> By "uncertain" knowledge, let me explain, I do not mean merely to distinguish what is known for certain from what is only probable. The game of roulette is

not subject, in this sense, to uncertainty. . . . Or, again, the expectation of life is only slightly uncertain. Even the weather is only moderately uncertain. The sense in which I am using the term is that in which the prospect of a European war is uncertain, or the price of copper and the rate of interest twenty years hence, or the obsolescence of a new invention, or the position of private wealth-owners in the social system in 1970. About these matters there is no scientific basis on which to form any calculable probability whatever. We simply do not know.[2]

Probability versus uncertainty: it doesn't sound like a very important distinction, but it was for Keynes.

All of rational choice theory is predicated on rational agents knowing the probabilities of all outcomes, the way we know all the risks—the posted odds—in a fair casino. The probabilities that rational agents attach to outcomes reflect their confidence in what will happen in the future. In the long run they will maximize their preferences to the extent that they are correct in their probability estimates of risky outcomes. Rational choice theory and well-behaved risk probabilities have been joined at the hip ever since a brilliant mathematician and physicist polymath, John von Neumann, showed that, in principle, we can calculate all of a rational agent's indefinitely many probability estimates by offering the agent choices between bets at different odds while holding the agent's preferences constant, and we can measure an agent's preference rankings by offering them choices while holding their risk probabilities constant.

Let's see why. Rational choice theory tells us that rational agents are omniscient with respect to all alternatives facing them. Since there are few sure things in anyone's future, what this really means is that they know the risks—the subjective probabilities, like casino betting odds—both for and against all the possible outcomes they have to choose among. RCT lets the economist uncover the rational agent's betting odds for all outcomes. If they use probability theory, we can cleverly show that rational agents get the probabilities right. Recall the exercise in revealed preferences: by offering a rational agent choices between all paired combinations of commodities and seeing which are chosen, the economist can build an agent's preference structure—the ranked list of their preferences. Once we've done this we can get agents to reveal their probability estimates for all alternative outcomes by offering them bets and seeing whether they take them or not. A simple example: We offer the RCT agent a bet of $2 if a coin comes up heads and $0 if it comes up tails or a straight payoff of $1. If the agent takes the bet, their probability estimate for heads is above 50 percent. Now we can start lowering the straight payoff to ninety-nine cents. If agents still takes the bet, we know their probability estimate for heads is even higher. By how much? We keep offering bets versus straight payoffs until the agent is indifferent between the bet and the payoff. The odds then give the probability that the rational agent attaches to the outcome. Since rational agents know all alternative outcomes and can rank them by preference, their beliefs about the future—their expectations—are as complete as these preferences. But unlike preferences, expectations are *cardinal*—they come in amounts between 1 (certainty an outcome will occur) and 0 (certainty that it won't). Von Neumann died before he could be awarded a Nobel Prize for showing this, but he achieved undying fame among economists nonetheless.

The rational agent lives in a casino with odds that can be figured out, always choosing the option that provides the highest *expected value*. Keynes called it *mathematical expectation*, though it is now more usually called *expected utility* (even though there is no such thing as utility): the product of its probability and its value to the bettor. One unit of expected value equals a 50 percent chance of winning two units of something the rational gambler wants. Rational agents maximize their expected utilities. They are rational gamblers.

Expected utility theory showed that RCT didn't require or presuppose certainty about alternatives facing rational agents. When risks can be rationally calculated, rationality does very nicely with probabilities. Once von Neumann showed how to construct expected utilities, maximizing them became the benchmark of rationality in a probabilistic world.

Keynes recognized that RCT required well-behaved probabilities, like the ones in a casino. But he argued that in the real world there is no reason to suppose such probabilities are pervasive. Economists and those they advise simply help themselves to the assumption that every outcome has a discoverable probability. He wrote, "In practice we [economists and business people] have tacitly agreed, as a rule, to fall back on what is, in truth, a *convention*."[3] A convention is not a hypothesis, an inductive generalization, or a well-supported belief about nature. It's just an agreement, implicit or explicit, among several parties to proceed in a certain way.

> The essence of this convention—though it does not, of course, work out quite so simply—lies in assuming that the existing state of affairs will continue indefinitely, except in so far as we have specific reasons to expect a change. This does not mean that we really believe that the existing state of affairs will continue

indefinitely. We know from extensive experience that this is most unlikely. The actual results of an investment over a long term of years very seldom agree with the initial expectation. Nor can we rationalize our behavior by arguing that to a man in a state of ignorance errors in either direction are equally probable, so that there remains a mean actuarial expectation based on equi-probabilities. For it can easily be shown that the assumption of arithmetically equal probabilities based on a state of ignorance leads to absurdities.[4]

Keynes pulled the rug out from under RCT when he argued that people often face situations of complete uncertainty—future outcomes to which no probability numbers can be attached. These are outcomes that rational choice is powerless to help us prepare for or choose among. The domain of uncertainty has a boundary that is hard to identify but that rational choice cannot cross.

The existence and extent of uncertainty is a major problem for economics. The insight about how preferences and probability estimates are related, from von Neumann, finds its way into every nook and cranny of economics. Where it can't get established because there are no probability numbers—the domain of uncertainty—economic theory is helpless.

This is one place where Keynes thought money comes in. Keynes argued that we always hold some money instead of less-liquid interest-bearing claims on future consumption (e.g., bonds or shares of stock) because we are all in the domain of uncertainty with respect to some, maybe much, of the future. *Precautionary balances* are liquid funds we keep on hand for rainy days—recessions, depressions, unemployment, ill health, or bad weather. *Speculative balances* are funds we keep on hand to jump on opportunities when things are looking up in the economy. Hanging on to more money than you need for short-term purchases can't be irrational when uncertainly rules and RCT can't help. Uncertainty, the absence of probabilities, thus gives money a real role in the economy, as precautionary or speculative.

Besides uncertainty, there is also a kind of money illusion operating in the real economy. People care a lot about money prices, not just relative (i.e., "real") ones. Money wages—the price of labor in currency—is a powerful example. People attach importance to their money-wage rate, not just to their real (inflation-adjusted) income. As a result the price of labor is *sticky*: it doesn't change as the supply and demand for labor do. Neither employees nor employers want to cut money-price wage rates when demand for the firm's products temporarily weakens. Workers don't want to earn less, and they think about *more* and *less* in terms of the money

Money—Who Needs It? Not Economic Theory! 93

price of their labor, not real price or the money price adjusted for inflation. Employers don't want to lose the experienced employees they have and then have to replace them with new, less experienced workers if demand bounces back quickly.

The "sticky wages" money illusion was just the tip of an iceberg that Keynes's followers thought they had discovered—one that revealed the power of money to shift an economy, the thing RCT forbade. In a paper published in 1958, "The Relation between Unemployment and the Rate of Change of Money Wage Rates in the United Kingdom, 1861–1957," William Phillips drew attention to the fact that over a long period of time, changes in the rate that money wages increased and decreased in the British economy correlated closely with changes in the rate of employment. The rates of change went up together and down together, even though the neutrality of money RCT imposes on the economy forbade this from happening.

Keynesian economists didn't care. They expanded this statistical finding to assert that changes in the employment rate correlated with changes in the rate of increase in all money prices, not just the price of labor. They called it the Phillips curve (see figure 5.3).

Keynes's work on the causal role of the amount of money in the economy—the speculative and precautionary demand for it, and people's

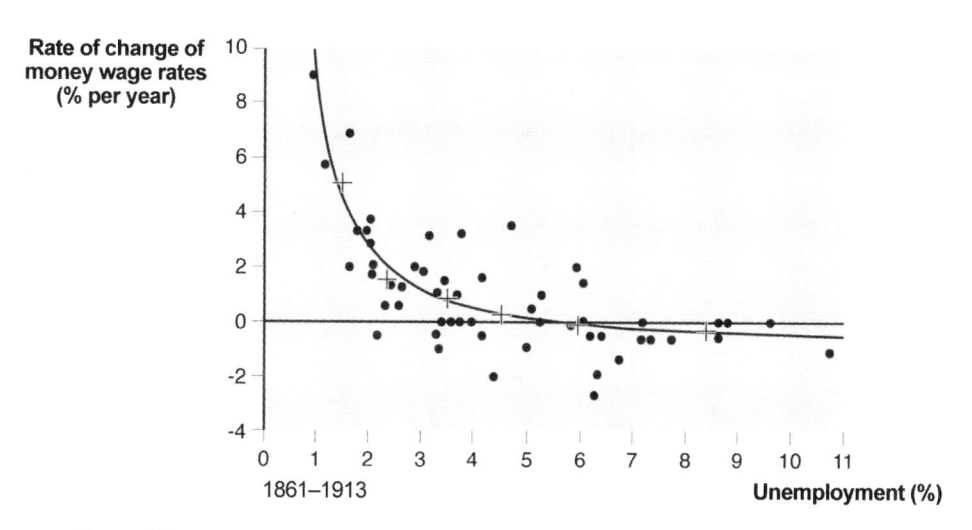

Figure 5.3
The Phillips curve for 1861–1913.

focus on money prices, especially the wage rate, instead of real prices—provided a ready explanation of the Phillips curve. That in turn explained the persistence of economic depressions as an effect of money, its velocity, and the price level. In a depression, deflation—moving down the *y*-axis on the Phillips curve—will push unemployment higher (rightward on the *x*-axis) because money saved as a precaution will have to reduce either the price level, *P*, or the total quantity of stuff the economy makes, *Q*. If prices, in this case wage rates, are sticky, only *Q* will fall, and with it will fall the amount paid to labor.

Keynes's followers converted his explanation for the decade-long depression of the 1930s into a policy tool that produced the steady postwar expansion of the '50s, '60s, and '70s. The stylized facts illustrated in figure 5.4 show how it can be done. The Phillips curve is a downward-sloping line. Raising the rate of inflation will reduce the level of unemployment. Calibrating how much to increase inflation and how much unemployment will be reduced is, like all matters of the real economy, beyond the scope of the stylized facts or their theoretical explanation.

But if money has a causal role in the economy, then we need to entirely rethink our understanding of the reality of rational choice and to refound economic theory on a different, more realistic model of human behavior. But such a model can't give us the beautiful first theorem of welfare economics.

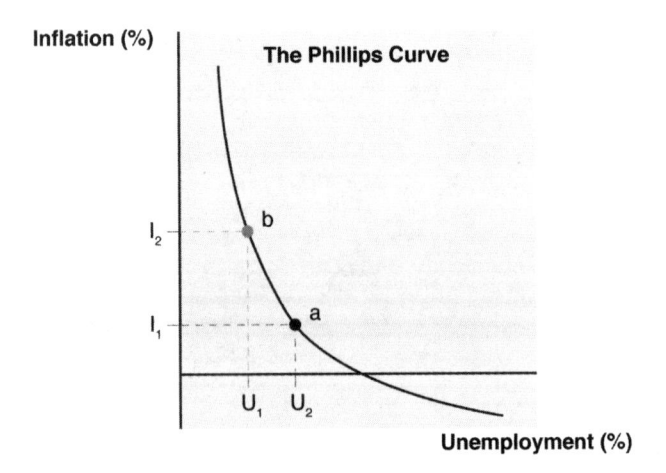

Figure 5.4
Using money prices to effect real changes in the economy.

It's easy to see why Keynes has often been identified as the first behavioral economist. He took seriously his obligation to at least sketch an alternative model of choice that would be consistent with his theory that the economy isn't always in or moving toward Pareto optimal general equilibrium. He famously resuscitated an ancient expression, "animal spirits," to describe departures from RCT rationality as important drivers of economic change. And he was the first to connect these departures from RCT with the failure of preferences and probabilities to mesh in the way RCT requires. More than once he decried the hypothesis that people make decisions by computing probabilities in risk-neutral ways. "Most, probably, of our decisions . . . can only be taken as a result of animal spirits—of a spontaneous urge to action rather than inaction, and not as the outcome of a *weighted average of quantified benefits*. . . . [I]f the animal spirits are dimmed and the spontaneous optimism falters, leaving us to depend on nothing but a *mathematical expectation*, enterprise will fade and die;—though fears of loss may have a basis no more reasonable than hopes of profit had before."[5] Actually inserting a behavioral economic revision of RCT into economic theory was no less daunting a task for Keynes than for the behavioral economists who followed him.

(Way) Back to the Drawing Board

Keynesian planning was too good to last. The method for fine-tuning the economy by varying the inflation rate to increase (but never to reduce) employment was already in trouble by the 1960s. That was when a British MP (who was to become chancellor of the exchequer) coined the term *stagflation*—a combination of high inflation and increasing unemployment. It was a combination that Keynes's theory, with its reliance on the causal power of money, couldn't address.

In an empirical science the succession of explanatory and predictive models usually shows a certain pattern or sequence, but there are some patterns that it never shows. Think back to the ideal gas law and the billiard ball model that explains it. As the technology of measurement improved and as new gases were isolated for experiment, the data on how pressure, temperature, and volume varied began to depart from the results of $PV=kT$. Changes had to be made in the equation and in the model of molecular interaction from which it was derived. The result was a sequence of models of increasing accuracy and range built up over two hundred years or so until we ended up with a quantum theory of gas behavior. But there was never a return to an older, already discredited model.

The same process happened in the study of genetics, though there the revisions were much more radical than in physics. Starting out with Mendel's hereditary factors, then substituting the label *gene*, biology refined the models over a century until Mendel's models had disappeared. In their place arose myriad different ways of dividing up DNA sequences to model more accurately how heredity is controlled. But no one looked back to a superseded model to explain data that a later model couldn't. This sequence of corrigible but improvable and sometimes replaceable models encapsulates the progress of science.

So what happened in economics? By the late nineteenth century, marginalist economic theory had become increasingly attached to the inertness of money in the real economy. Just before the Great Depression Friedrich Hayek labeled it "the neutrality of money." Economists recognized that RCT gave us the Pareto optimality of general equilibrium and in doing so reflected the absence of any money illusions on the part of economic agents. The theory dominant during this period came to be called *classical economics*.

But the model couldn't deal with the decade-long slump after World War I or the Great Depression that followed. Many economists, following Keynes, framed a new model that sought to explain both the general equilibrium conditions that classical economics described and the multiple nonoptimal equilibriums into which an economy can drive itself—long periods that few would call optimal or efficient. Keynes's students and other economists, building on *The General Theory of Employment, Interest, and Money*, developed a set of models that they called *neoclassical economics*, which they thought incorporated the insights gained in the nineteenth century and those from Keynes's theory. Over the postwar period leading up to the 1970s, the models were used by governments across developed economies. And then neoclassical economics—the Keynesian models—stopped working.

How did economic theorists respond to this surprising, disturbing, disappointing turn of events in their science? Here's a famous passage by two Nobel Prize winners we'll have occasion to quote more fully in the next chapter:

> We dwell on [the] halcyon days of Keynesian economics because without conscious effort they are difficult to recall today. . . .
> That [the] predictions [of Keynesian theory] were wildly incorrect and that the doctrine on which they were based is fundamentally flawed are now simple matters of fact involving no novelties in economic theory. The task now facing contemporary students of the business cycle is to sort through the wreckage.[6]

The result of this sorting through wreckage and the consequent so-called advance in economic modeling was not a new and better model of the economy. Instead, it was something that proudly came to call itself "new classical" economics. But this was a misnomer, since the only thing really new about it was the label "new." It was the old wine of classical economics in new bottles of post-Keynesian laissez-faire. Economic theory went back, not forward, to the status quo ante. New classical economics reimposed "the discipline"—Robert Lucas and Thomas Sargent's words—of markets that completely clear and agents who fully optimize.

One of the new classical theory's most sternly proclaimed tenets was the neutrality of money, for exactly the same reasons classical economics was committed to it: people are rational, and rational agents are interested only in real prices. There is no room for money in microeconomics, and no room for it in macroeconomics, either. Lucas entitled his Nobel Prize lecture "Monetary Neutrality." *Hamlet* without the prince.

In the next chapter we'll consider how RCT's writ expanded beyond microeconomics, the theory of individual economic choice, into macroeconomics, the theory of aggregate economic forces, and then altogether erased the whole compartment of the discipline of economics that Keynes had built. Like *utility*, the word *macroeconomics* hung around, but the thing itself disappeared.

6   Back to the Future with New Classical Macroeconomics

Contemporary macroeconomic theory is almost entirely a construction of pure reason, running thought experiments on rational agents unaided by empirical evidence about what people actually do. As we'll see, its starting point is the collapse of Keynesian economics in the late 1970s, and its finish line is the economist's favorite version of Adam Smith's *Wealth of Nations*.

To experience the back-to-the-future reasoning for ourselves, we have to walk through some equations and graphs. We won't have to do calculations the way empirical scientists do, and the few graphs we'll look at don't have numbers on their axes. The equations are all just elementary algebra. They look formidable because they use Greek letters and subscripts. But there isn't so much as an exponent—a square or cube or even a square root, let alone a *dx/dy* or a ∫ *dx*. You don't need more than ninth-grade math to read them and understand what they say. You won't need to remember them. And we'll put them into words. They'll be child's play after the math in chapter 4—the constrained maximization problem of the representative household. One reason we had a look at it back there, however, was its role in motivating the simpler macro models this chapter reviews. Along the way there will be some more history that helps us understand the equations. Most importantly, we'll work through the reasoning that starts with RCT and produces modern macroeconomics with almost no help from data about the real economy.

The Rip Van Winkle Effect in Economics

Here's a potted history of macroeconomics that picks up the story where we left it at the end of the previous chapter. There was no such thing as macroeconomics until the 1930s, when John Maynard Keynes began trying to figure out why the Great Depression happened. Until then there was only microeconomics. Fifty years later, when Keynesian models stopped working, macroeconomics disappeared, leaving just the label. Only microeconomics was left, and economics was back to where it was before Keynes.

There is more to micro theory than proving the existence of the Pareto optimal general equilibrium. But everything else rests firmly on RCT, convex indifference curves, and the proof of the first theorem. Microeconomics is what can be built up from individual decision-making processes, added up and aggregated to become markets and industries, national economies, and international trade. Everything that happens can be analyzed down to the actions and reactions of individual firms and individual consumers.

Treating the models of microeconomics as reasonably accurate descriptions of what was going on in real economies became much more difficult after the onset of the great worldwide depression of the 1930s—"the slump," as the British called it. It was evident that economies were nowhere near Pareto optimal general equilibriums, and they didn't seem to be headed in that direction, either. There were vast surpluses, gluts, especially of agricultural products and labor—lots of people willing to work at almost any wage, and no demand for them at any price. Markets weren't clearing. Things were so bad for so long that it looked like the economy was in a stable, long-term general equilibrium that was utterly inefficient and certainly not Pareto optimal. Classical theory said that couldn't happen.

Keynes set out to explain why a situation that microeconomics deemed impossible had emerged. Instead of starting with individual choice behavior and arithmetically building up markets, industries, and economies, he began by looking at the whole economy from the top down without worrying about individual microeconomic pieces. Keynes began to ponder the relationship between the total supply of money, the interest rate, the unemployment rate, and other "macro" variables (ones that characterize the whole economy). The result was his book *The General Theory of Employment, Interest and Money*, which established macroeconomics as a distinct subdiscipline with its own models, not built out of or mathematically derived from microeconomic models of the economy.

The macroeconomic models Keynes inspired were ones that allowed individual markets and whole economies to enter and remain for long periods in highly inefficient, Pareto nonoptimal equilibriums. Classical microeconomic models treated the business cycle—booms and busts—as local and temporary. They were short-term departures from the normality of market-clearing general equilibrium that would correct themselves by price movements.

Keynes showed how, in principle, markets could get stuck far from efficiency. He focused particularly on lack of demand as the obstacle to the economy's return to conditions that microeconomics deemed normal. Weak demand for labor, for loans to invest in new business, and for money to purchase consumer goods would produce persistent business slumps. Once demand for these three categories fall, expectations of further near-term declines in demand are self-fulfilling prophesies as people increase savings to prepare for a prolonged slump—a suboptimal equilibrium. Two features of Keynes's theory departed from "classical," pre-Keynesian microeconomic theory: First, Keynes was not interested in deriving his theory from RCT. As noted in chapter 5, he didn't even accept RCT. If anything he was a behavioral economist avant la lettre, treating people as driven by their gut feelings, or "animal spirits," as he called them. More importantly, he reckoned that economic decisions are made under conditions of radical *uncertainty*, not omniscience. Without well-behaved probabilities, RCT choosers can't start acting rationally at all.

Keynes's students spent the decades after the publication of *The General Theory* developing mathematical models of the macroeconomy. These were sets of equations that related macro variables like the money supply, the interest rate, the rate of investment, and the rate of savings. These models revealed the possibility of what classical microeconomics deemed impossible: prolonged, persistent depressions. They also suggested ways for governments to shorten such slumps instead of waiting for markets to eventually self-correct, as classical microeconomic theory hoped they would. The means were to increase the demand for goods and labor in the whole economy: increase the money supply, lower interest rates, spend more in government purchasing than the government receives in taxes (and borrow the difference), and increase the rate of inflation. Recall the Phillips curve discussed in chapter 5: increase the rate of inflation to lower unemployment.

The Keynesian models were embarrassingly simple; they were hardly even algebra. You can't find them in most of the new classical macroeconomics textbooks anymore. Like microeconomic models, Keynesian models explained stylized facts—trends and qualitative regularities that economists found obvious in the macroeconomic data that government agencies had begun to collect.

Doing more than deriving stylized facts from idealized models requires that we determine the mathematical values of the variables in the models. Keynesian economists turned their attention to this task. They hoped to start the iterated process of model refinement, data collection, model correction, and more data collection that characterizes an empirical science. Even outsiders began to hope that economics was finally following a trajectory of development familiar from the natural sciences. Macroeconomics spawned *econometrics*—the employment of statistical techniques on data to quantify macroeconomic variables and make educated guesses about causal relationships among them in order to test macroeconomic models. There was optimism that Keynesian theory was more or less correct and was on track to make further refinements. Governments across the developed world began employing its models to "fine-tune" their economies. They sought to mitigate recessions and prevent depressions by stimulating demand, using inflation to reduce unemployment, and keeping the economy on an even keel by varying interest rates and the money supply. These were all things classical microeconomics said weren't needed and in fact couldn't be done in an economy of rational agents.

Alas, it was too good to be true. Keynesian macroeconomics unraveled in the 1970s. The models stopped working, or at least the phenomena that the models ruled out—ever-increasing inflation coupled with higher unemployment—became widespread and persistent. Macroeconomists had to go back to the drawing board, looking for what Keynesian models had gotten wrong. The conclusion that became widespread in the profession was: *pretty much everything*. In the famous paper quoted at the end of chapter 5, "After Keynesian Macroeconomics," published in 1979, two future Nobel Prize winners, Robert Lucas and Thomas Sargent, summed things up. Their prose is so purple that the passage is worth quoting in full.

> We dwell on [the] halcyon days of Keynesian economics because without conscious effort they are difficult to recall today. In the present decade, the U.S. economy has undergone its first major depression since the 1930s, to the accompaniment of inflation rates in excess of 10 percent per annum.... These events ... were accompanied by massive government budget deficits and high rates of monetary expansion, policies which ... promised according to modern Keynesian doctrine rapid real growth and low rates of unemployment.

That these predictions were wildly incorrect and that the doctrine on which they were based is fundamentally flawed are now simple matters of fact, involving no novelties in economic theory. The task now facing contemporary students of the business cycle is to sort through the wreckage.... Though it is far from clear what the outcome of this process will be, it is already evident that it will necessarily involve the reopening of basic issues in monetary economics which have been viewed since the thirties as "closed" and the reevaluation of every aspect of the institutional framework within which monetary and fiscal policy is formulated in the advanced countries.[1]

Going back to square one meant two things to the new classical macroeconomists. From that point forward, all macroeconomic models were to be "disciplined" by honoring two conditions, stated in four simple words: agents optimize; markets clear. In other words, macroeconomic models had to start from the same assumptions of rational choice theory as microeconomics, and reach the outcome the first theorem enjoined: Pareto optimal general equilibrium. One of the most influential new classical macroeconomists put it this way: "The basic [macro] model relies on two key elements for making predictions about the behaviour of quantities and prices in the real world. First, we use the model's *microfoundations*, which derive from the optimizing behavior of individuals subject to budget constraints. Second, we exploit the notion of market clearing, a process that reflects the efficient matching of potential buyers and sellers of goods, credit, labour services and so on."[2]

So where exactly had Keynesian macroeconomic theory gone wrong? Its mistake had been to assume that individual economic agents aren't as smart as economists. Just like economists, ordinary people eventually wise up to Keynes's macroeconomic theory and then act to prevent governments from using it to manipulate them. Keynes wrongly thought people suffer from all kinds of money illusions. His followers assumed that if governments inflated currency, people would just think they were richer and buy more. If governments ran deficits and lowered taxes, people would have more money in their pockets (from government purchases and higher post-tax take-home pay). They would spend more and increase demand, thus stimulating the economy. But if people are rational, like economists, eventually they'll see that *all* prices are rising, not just the ones they keep track of to make their economic decisions. At that point inflation will stop changing the pace of economic activity. Things will return to where they were before the inflation. The same is true for government deficits and tax cuts. People know that in the long run the government will have to spend less and raise taxes to pay off the debt resulting from short-term deficit spending and tax cuts. What will people do? They're rational. They will know that the reckoning is coming. They will cut spending and start saving in order to pay future higher taxes and to survive the government cuts that they know to expect.

New classical macroeconomics was founded on the supposition that every individual has rational expectations about the future. A rational expectation about the future is not one that simply assumes the future will be like the past, but one that incorporates enough economic theory to correct expectations when the government does something unexpected. This postulate—that people are not just rational but also understand economic theory—is at the heart of new classical economics. As we will see, it was the only thing macroeconomists thought they needed to square reality with the classical economics Keynes rejected.

We can illustrate how new classical theory uses rational expectations to unravel Keynes's theories by going back to an argument Milton Friedman tried to pick with Keynesians more than a decade before Lucas and Sargent's broadside. It was an argument Friedman thought he'd lost but it eventually inspired the anti-Keynesian counterrevolution.

Recall the Phillips curve from chapter 5. It plots real data from a half century that show how changes in inflation rates and unemployment rates were inversely correlated. Keynesian economists generalized from that relationship to establish an inverse relationship between changes in inflation rates and all increases or reductions in economic activity. Then they flipped it and turned it into a policy tool: to increase economic activity, increase the rate of inflation. While this policy was working well in the hands of the Kennedy administration in the United States, the Wilson government in the United Kingdom, and in certain European countries, Friedman noticed that it couldn't work if rational employers were smart enough to use the evidence before their eyes. He accepted the data that showed there might be a temporary Phillips curve. But he argued that it couldn't last.

Friedman helped himself to the same distinction between the short run and the long run that Keynes had ridiculed as too easy an out for economists and imposed it on the Phillips curve of changes in inflation rates versus unemployment. He reasoned that if employers don't suffer from a money illusion in the long term, inflation will have no impact on unemployment. The long-run Phillips curve will be vertical. But employers may, over a short period, miss the fact that all prices are rising, not just the prices they pay and charge. It's easy to do when your prices are the ones you're most interested in. If their selling prices increase, firms might hire more labor, at least temporarily. But eventually they'll realize that all prices are rising, not just the ones they charge—business has not improved. Then they will "let go" the extra workers they hired, bringing unemployment back to where it had been before the process began. It stands to reason, when expectations are rational and correctable! Friedman's reasoning is nicely illustrated in a diagram that became familiar to all new classical economists (figure 6.1). It shows how, in principle, the only path the unemployment rate can follow brings it right back to where it started.

Start at the inflation rate and unemployment rate of point A and move inflation up. Employers experience a temporary money illusion and think higher prices mean more demand, so they hire more workers, moving unemployment to the left along the short-run Phillips curve, reducing it. This brings the economy to point B, a lower level of unemployment at a higher inflation rate. But it can't last. Once employers see that business hasn't improved, that it's just a mirage of inflation, their short-run Phillips curve shifts unemployment to the right: they fire the recently hired workers. Unemployment goes back to the previous level, now at point C, where inflation is higher than it was at point A.

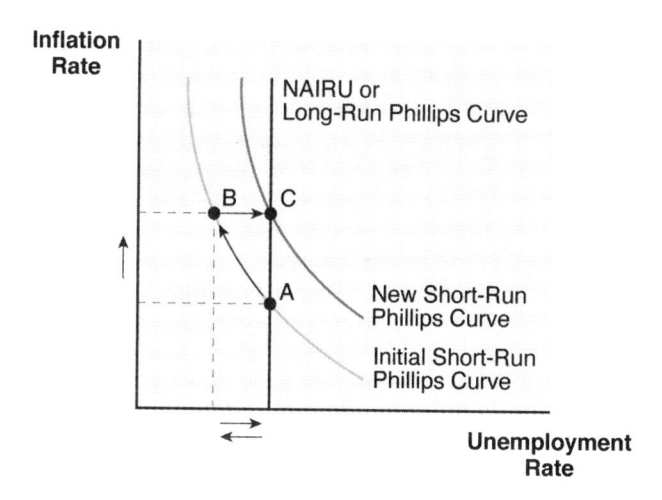

Figure 6.1

How short is the short run and how long is the long run? When Friedman published his paper, the short run had about another decade to go. Meanwhile, Friedman drew another conclusion from his diagram: the vertical line of the long-run Phillips curve is actually the level of *full employment* in the economy: a "natural" equilibrium level of unemployment that can't be reduced further. Full employment isn't zero unemployment. The unemployed will always be with us, even in the Pareto optimal general equilibrium. (We'll see in chapter 7 how this insight drove the new classical theory of the labor market as "friction.") This means we can't measure the long run as the length of time it takes the economy to get to zero unemployment. So what level is the natural level, the level that will tell us when the long term finally arrives, when, as Keynes wrote, "the storm is long past [and] the ocean is flat again"?[3] The answer is one we saw a few times in chapter 5. Economic theory calibrates time periods recursively: the long term is the period it takes to get to equilibrium, where unemployment is at its natural rate. How long does it take for the economy to achieve equilibrium? It takes the amount of time, whatever its duration, that it needs to reach equilibrium, to its natural rate. But how long *is* that?

Well before his Nobel Prize–winning insight about the natural rate of unemployment, Friedman advanced another influential claim that would later be used to further unravel Keynesian macro models and motivate new classical models. This was the *permanent income hypothesis*: that economic agents maximize lifetime utility subject to their expected lifetime earnings by borrowing when young, repaying debt in middle age, and living on savings in old age. Assuming agents are foresighted, long-term rational maximizers in their lifetime consumption patterns was compatible, Friedman noted, with macroeconomic data that the US savings rate remained constant while incomes were increasing. The Keynesian model wasn't compatible with it. According to Keynes's theory, there was a marginal propensity to save that increased with wealth: the richer people got, the more income they were supposed to save. This weakening growth in demand led to recessions or even economic depression. Friedman argued that the data didn't seem to bear this out. Equally important to Friedman was the *dynamic rationality* of agents who plan their lifetime consumption and savings because they can discriminate real-price changes from money-price changes. The permanent income hypothesis both repudiated the money illusion and attributed a long-term, lifetime planning horizon to everyone.

Perhaps it won't have escaped your notice that this is the same Milton Friedman who defended RCT by saying it wasn't to be taken literally as a claim about the cognitive activities of individual rational agents but was just a device for predicting downward-sloping demand curves. Fifteen years after his manifesto for the discipline, "The Methodology of Positive Economics," Friedman was taking RCT very seriously indeed. So much for revealed preferences.

New Classical Macroeconomics at Work

Having cleared away the wreckage, Lucas and Sargent began the project of rebuilding: "The hypothesis of *rational expectations* is being imposed here. Agents are assumed to make the best possible use of the limited information that they have and to know the pertinent objective probability distributions. This hypothesis is imposed by way of adhering to the tenets of equilibrium theory."[4] Generalizing from Friedman's insight about employers, new classical theory deals with all departures from equilibrium as temporary aberrations: agents sometimes develop incorrect expectations, and then the economy goes out of whack until their expectations are brought back in line with reality.

> The key step . . . [is] to relax [one] ancillary postulate used in much [pre-Keynesian] economic analysis that agents have perfect information. The new classical models still assume that markets clear and that agents optimize. . . . [E]ach agent is assumed to have limited information and to receive information about some prices more often than other prices. On the basis of their limited information . . . agents are assumed to make the best possible estimate of all the . . . prices that influence their supply and demand decisions.
>
> Because they do not have all of the information necessary to compute perfectly the relative prices they care about, agents make errors in estimating the pertinent . . . prices, errors that are unavoidable given their limited information. . . . [U]nder certain conditions, agents tend . . . to mistake a general increase in all . . . prices as an increase in the . . . price of the good they are selling, leading them to increase their supply of that good over what they had previously planned. Since on average everyone is making the same mistake, aggregate output rises above what it would have been. This increase of output above what it would have been occurs whenever this period's average economy-wide price level is above what agents had expected it to be on the basis of previous information. Symmetrically, aggregate output decreases whenever the aggregate price turns out to be lower than agents had expected.[5]

Because agents are rational and modify their expectations as evidence comes in, they correct their errors, bringing aggregate demand and supply back toward equilibrium. The microfoundation of macroeconomics, that buyers and sellers are rational and correct their errors about price levels, helped new classical economists explain why Keynes's theory failed in the 1970s. *Microfoundation* is a buzzword in macroeconomic theory. The new classical demand for microfoundations will come up again and again.

All these economists needed to make macroeconomics just a branch of microeconomics was rational expectations: "We apply this same economic science to understanding the workings of the overall economy—that is, to study real GNP [gross national product], employment and unemployment, the general price level and inflation, the wage rate, the interest rate and so on."[6] It's as if economic theory had been knocked senseless in 1930 and woke up fifty years later, having forgotten nothing from before it had been hit on the head but having learned almost nothing from the blow.

The Central Banker's Model of the Macroeconomy

Macroeconomic theory is dominated by models that reflect new classical economists' focus on short-run expectations and long-run natural rate equilibriums. These are *dynamic stochastic general equilibrium* (DSGE) models. They are dynamic because the models describe the economy over more than one time period, stochastic because they describe how the market reacts to outside events or shocks that economic agents can't predict, and are about general equilibrium because the models are designed to show how the market moves from shocks back to the Pareto optimal, market-clearing outcome required by the first theorem of welfare economics. These models describe the economic variables that policymakers care about: the total output of the economy, the interest rate, the rate of inflation, and central bank operations that affect the money supply.

It's obvious why economists want mathematical models. If, as in empirical science, the equations of the model reflect economic reality, the model can be used to explain, predict, and modify the economy. Increase or decrease one of the variables symbolized in the equations and see how that changes the values of the other variables. Feed data for some of the variables into the model and use the equations to compute other variables. Check your prediction against reality. Revise the model and repeat. If we are going to figure out what's happening in economic theory now, we need to take apart one of these models.

DSGE models look impressive, but they are not mathematically complicated. We'll explore a model with five equations that will show how rational but fallible expectations are packed into them, what a difference expectations make when they are temporarily wrong, and how the economy corrects them. The simplified DSGE model we'll walk through comes from a textbook, Gregory Mankiw's *Macroeconomics*. In the text it's pretty clear how a DSGE model is developed. It's not a matter of extracting equations from data. It's more like a thought experiment in which you add stuff to equations by thinking through how economic forces would change your decisions if you were a rational agent.

The first equation looks like this:

$Y = \bar{Y}_t - \alpha(r_t - \rho) + \varepsilon_t$

It says that Y, total actual output demanded in year t, is going to be different from the market-clearing level of output, $\bar{Y}_t$, and the difference will be a matter of the rest of the equation: $\alpha(r_t - \rho)$ and ε_t.

The stuff in the parenthesis of the first of these two, $(r_t - \rho)$, is the discrepancy between the real interest rate, r_t, operating in that year, and the natural, long-term interest rate, ρ, that would operate if the economy were at its equilibrium. We've seen the term *natural* before, in Friedman's claim that there is a natural rate of unemployment that is the nonzero rate of unemployment that an economy produces when working at full employment. This is the same level at which the real interest rate equals its natural rate. When the actual interest rate is higher than the natural interest rate, the economy's total output will shrink to below the natural rate because consumers are saving too much and businesses are investing too little. When the actual interest rate is lower than the natural rate, the economy will produce more than it would at the natural rate because consumers are spending more and saving less, and firms are borrowing to buy inputs and build factories. Alpha, α, is the coefficient that weights the effect on output Y of the difference between real and natural rates of interest, $(r_t - \rho)$. Alpha will be large if the economy's output is very sensitive to the level of savings or investment and small if it isn't.

At the end of the equation, ε_t is an outside, exogenous variable, like increased or decreased government spending, a tax cut no one expected, or a change in people's level of optimism or pessimism about the economy's future that changes their willingness to buy or save. Economists call this a *random shock* to the level of output. *Random* here just means unforeseeable by rational agents inside the model. It's what makes the model stochastic.

The first equation in the model puts all of that in symbols: $Y = \bar{Y}_t - \alpha(r_t - \rho) + \varepsilon_t$. In words, Y, total actual output demanded in year t, is going to be different from the market-clearing level of output, $\bar{Y}_t$. By how much? Actual output demanded depends on the market-clearing general equilibrium level of output, $\bar{Y}_t$, but is reduced by some amount, the coefficient α, times the difference between that year's real interest rate, r_t, and the natural rate of interest, ρ, plus the value of a random or stochastic shock, ε_t. In case the words don't help, the ideas behind the equation are simple. The economy is out of whack when the real interest rate departs from its natural, long-term, market-clearing level.

One line of math, no calculus, no quadratic equation, just some Greek letters. We'll see some of the same notation in the other equations.

The second equation in the model is named after Irving Fisher, the most important American economist of the classical, pre-Keynesian period. It is

$r_t = i_t - E\,\pi_{t+1}$.

It sets the "real" interest rate people are paying in a given year, r_t, equal to that year's nominal money interest rate, i_t, minus the next year's expected rate of inflation, $E\,\pi_{t+1}$: E for expected and π_{t+1} for inflation in year $t+1$. The *real interest rate* is "corrected" for expected inflation or deflation, so it can be compared from year to year without adjustment for fluctuations in money prices. The *nominal interest rate* is paid in money and is therefore affected by inflation and deflation. The real interest rate that people are paying equals the money interest rate corrected for the rate of inflation or deflation they expect to prevail in the next year. This is the first point at which expectations come into the model.

The Fisher equation tells us that if people are temporarily wrong in their expectations about inflation, they will accept a money interest rate that temporarily pushes the real interest rate too high or too low. Since r_t is also in the first equation above, changing it pushes the economy's output away from its natural rate.

The third equation is called the Phillips curve equation, not because it descends from Phillips's data, but because it reflects the way inflation results from expected inflation and the economy's output level. It is

$$\pi_t = E_{t-1}\pi_t + \phi(Y - \bar{Y}_t) + v_t.$$

It expresses the relationship between this year's inflation, π_t, past expectations about it, $E_{t-1}\pi_t$, and the discrepancy between this year's actual output and the natural rate of output, $Y - \bar{Y}_t$. The equation also adds another stochastic variable, v_t, to symbolize external shocks to the supply of stuff—unpredicted (and unpredictable) shocks like an oil embargo, a war, the collapse of the housing market, a pandemic.

Here's how this equation for the inflation rate, π_t, gets put together. This year's short-term total output of the economy is going to be either more or less than the long-term, "natural" output level at market-clearing equilibrium. If it's more, then that difference from the natural level is going to cause inflation because the economy is producing more than its long-term sustainable capacity. So this year's actual rate of inflation will reflect the difference between actual output and natural output, plus the rate of inflation we expected last year for this year. In the Phillips curve equation, ϕ works like α in the first equation. It's a coefficient that measures the sensitivity of the real rate of inflation to the gap between actual output and market-clearing optimal output.

To reiterate: the rate of inflation of year t, π_t, is determined by the rate of inflation in year t that was expected in the previous year, $E_{t-1}\pi_t$, plus some multiple, ϕ, of the difference between the total output and the natural rate of output for year t, $\bar{Y}_t$, plus another random supply shock, v_t. Notice that if people predict the next year's actual inflation accurately, then the discrepancy between the actual output and the natural level of output must be zero and there are no supply shocks. If everyone knows exactly how much more inflation is coming, expectations will have no effect on the economy—there will be no money illusion. It's only when inflation is a surprise that there will be a short-term change in output. That's Friedman's point about the natural rate of unemployment, now generalized to the natural state of the whole economy. Inflation that people mistake for an increase in demand or in what buyers are willing to pay leads to temporary increases in production (perhaps by hiring more workers), but production decreases again as erroneous expectations come into line with reality.

The fourth equation tells us how economic agents project the next year's rate of inflation based on the current year's rate. In the simplest case the expected inflation rate for year $t+1$, $E_t\pi_{t+1}$, is the same as the actual rate of inflation in year t, π_t:

$E_t \pi_{t+1} = \pi_t$.

If agents are rational and well informed, in the long term they will accurately predict inflation. When that happens, we can set expected inflation, $E_t \pi_{t+1}$, equal to actual inflation, π_{t+1}. (Notice how the time subscripts for $E_t \pi_{t+1}$ and π_t are different.)

The last equation of the five in the model is the most complicated and hardest to explain. It's an equation that describes how central banks—the Federal Reserve Bank in the United States (familiarly called the Fed), the Bank of England, the European Central Bank—respond to inflation and deflation. Rates of inflation or deflation are signals that the economy is producing more or less than the natural rate of output. The central bank wants to move interest rates up and down so that the whole economy's actual output equals its natural rate of output at general equilibrium. At that point there is no inflation because there is no discrepancy between nominal (money) prices and "real" prices. Here's the equation:

$i_t = \pi_t + \rho + \theta_\pi (\pi_t - \pi_t^*) + \theta_y (Y - \bar{Y}_t)$.

It's about what determines the nominal money-price interest rate, i_t, in a given year. The central bank will use this equation to set a nominal interest rate that moves the economy to its general equilibrium as fast as it painlessly can.

We've seen several of these elements before: the variables in parentheses and the coefficients in front of them. As in the other equations, the coefficients measure the effects of the differences inside the parenthesis. The nominal rate of interest in year t, i_t, is determined by inflation in year t, π_t, plus the natural rate of interest, ρ, plus the two differences, each multiplied by a coefficient, θ. The first set of parentheses represents the difference between real inflation in year t, π_t, minus the central bank's target for real inflation in year t, π_t^*. The second pair has another multiple, θ_y, in front of it. It's the difference between actual output in year t and the natural rate of output. The last term on the right, $(Y_t - \bar{Y}_t)$, is the difference between actual output and the equilibrium level of output. We saw it in the third equation. If positive, it's going to produce price inflation. How much inflation it creates depends on the coefficient θ_y. The term $(\pi_t - \pi_t^*)$ is the difference between real inflation in year t and the central bank's target for inflation that year. If that difference is positive and large it will affect the nominal interest rate by an amount that depends on the coefficient θ_π.

Combining the five equations in the model gives us four things: the total level of output of the economy, the real interest rate, the rate of inflation, and the money rate of interest. Four of the equations reflect the interplay between the actual values of key economic variables and rational but fallible expectations about these values. With these equations we can see how a one-time exogenous shock to the economy can change the paths these variables take. We'll also be able to appreciate how fully new classical macroeconomics rolls economic theory back to Smith's time.

The first and most important thing theorists do with the model is describe the forms that its equations take under long-run equilibrium. This is the state of the economy in which all changeable variables are equal to what the model identifies as their natural values. These are the values that obtain when, as Mankiw describes it, the market is in the state "around which the economy fluctuates."[7] When the market is in its natural state the complicated equations we just walked through reduce themselves to four, as the short-term mistaken expectations drop out, and everyone knows everything about the economy's future in its steady state.

In long-run equilibrium the five equations simplify to these:

$Y = \bar{Y}_t$; the actual output of the economy is at its natural rate of output.

$r_t = \rho$; the real rate of interest is at its natural rate.

$E_t \pi_{t+1} = \pi_t = \pi_t^*$; the expected rate of inflation for the next year is exactly the same as the actual rate of inflation in the current year and is equal to the central bank's planned inflation rate for the current year.

$i_t = \rho + \pi_t^*$; the nominal interested rate in the present year is exactly equal to the natural rate of interest plus the central bank's planned inflation rate.

At these values, when the economy is in long-run equilibrium, the real prices of every commodity make the supply of every good or service on every market, including the labor market, exactly equal to its demand; there is no excess demand. In consequence, at the natural rate of output and interest, there is no shortage and no surplus on any market. Again, this includes the labor market. There is zero unemployment. All potential suppliers of labor sell all they want at the market price. Anyone not working just doesn't want to work at the prevailing wage rate. There is no involuntary unemployment.

It's worth noting again that when the equations reduce to these long-run equilibrium values, money no longer plays any role in the economy. No one holds money balances beyond what they need in day-to-day exchanges for shopping, paying wages, and so on. All unspent earnings are invested at the real interest rate, and neither money-price inflation nor changes in its rate have any effect on the economy. This feature of the long run reflects what economists call the *classical dichotomy* between the real forces that drive the economy and the nominal factors, such as the supply of money, prices, and interest rates, denominated in units of money. The rates of inflation in these prices, which appear to be causes of economic activity, play no role in the long run. Money is neutral!

There's Got to Be a Natural State; Microfoundations Tell Us So

What guarantees that the economy is ever in its natural state, or even that it has one at all? This is where new classical macroeconomics' insistence on microfoundations for any macro model come in. The existence of a natural state for the economy is not inferred from data. It's a deduction derived from microeconomic assumptions—more models, ones we've seen before, in the proof of the first theorem of welfare economics.

Two models are particularly relevant: one for households, the other for firms. The two models begin with one advanced originally by an important twentieth-century philosopher, Frank Ramsey, who was working on economics as something of a pastime or distraction from his work in philosophy.[8] Ramsey was a socialist and in 1926 set himself the puzzle of how a single planner could address the problem of choosing a socially optimum distribution of consumption and savings for an entire economy. About forty years later his paper was rediscovered by influential mathematical economists Tjalling Koopmans and David Cass, who extended it in a number of ways. In particular they first substituted for Ramsey's single planner a single representative agent, a Robinson Crusoe, deciding how to divide his time between production and leisure to maximize his utility. Mathematical modelers work with two sets: first, a set of indefinitely many identical households, and second, an equally large set of firms. The indefinitely many identical households in the first set have identically shaped indifference curves and all choose patterns of consumption and savings over time that maximize lifetime utility. In the other set, equally large, firms pay for labor and capital inputs at their marginal productivity and sell their output in a perfectly competitive market.

The households grow at a certain rate and rent their initial capital to firms. They divide their incomes, from labor and capital, between immediate consumption and savings depending on the real interest rate, ρ, in an indefinite number of periods so as to maximize lifetime utility. The higher the real interest rate, the more households save in that current period. Households' utility maximization is constrained by their budgets: the present value of future lifetime consumption cannot exceed the sum of a household's initial wealth and the present value of its lifetime wage income.[9] (Sounds like Friedman's permanent income hypothesis, doesn't it?)

Mathematical economists were able to prove that an economy composed of households and firms that satisfy these Ramsey-Koopmans-Cass models is at a general equilibrium of the sort new classical economists describe as the long-run, natural, normal state of the DSGE model. In chapter 4 we looked at some postcards from a part of the derivation in David Romer's *Advanced Macroeconomics* textbook—the household's intertemporal constrained optimization problem and the Lagrangian that solves it.

Why care about this equilibrium? Why task central banks with trying to achieve it? The answer is one we're familiar with from chapter 3. As Romer writes:

> A natural question is whether the equilibrium of this economy represents a desirable outcome. The answer to this question is simple. The *first welfare theorem*, from microeconomics, tells us that if markets are competitive and complete . . . then the decentralized equilibrium is Pareto-efficient. . . . Since the conditions of the first welfare theorem hold in our model, the equilibrium must be Pareto-efficient. And since all households have the same utility, this means that the decentralized equilibrium produces the highest possible utility.[10]

DSGE models are logically derived from an economy composed of utility-maximizing households and competitive firms of the sort familiar from microeconomics. A world the DSGE models describe correctly is one in which the economy has a natural state and in the long run is always there, in a Pareto-efficient general equilibrium. Thus the DSGE macro model is, as the new classical economists say, *micro-founded*.

What about the short run, while we're all still alive? In the short run the economy is always on a round trip back to general equilibrium (even if it never really gets there).

Beginning with the simple four-equation model of long-term equilibrium, we can build up to the five equations of the bigger DSGE model we started with. They are all supposed to reflect ways in which the short-run conditions of the macroeconomy can depart from the long-run general equilibrium and to identify the forces that move it back toward that blessed state.

How do macroeconomists get from the simple equation $Y = \bar{Y}_t$ back to $Y = \bar{Y}_t - \alpha(r_t - \rho) + \varepsilon_t$? By turning to thought experiments in the *microfoundations*. That is, the more complicated short-term equation is logically derived from microeconomic theory by combining the RCT model with the rational but fallible expectations of agents treating money-price changes as signals of real-price changes. That's because it's all they have to go on, in the short run. Macroeconomists put themselves in the shoes and minds of rational agents and ask themselves how they would respond to the signals about real economic variables sent by changes in the monetary value of these variables. Like the economists who posited rational agents, none of these agents will be subject to money illusions, but none of them have a source of information about real values besides money prices. Plugging the Fisher equation for the real interest rate into the equation above for total output makes the role of money prices explicit: Y is going to depend on $E\,\pi_{t+1}$, the rate of inflation people expect next year.

Sometimes people underestimate the next year's inflation and so the real interest rate is temporarily greater than the natural rate of interest, the value that will prevail in the long run. Because people mistakenly think interest rates are rising, rational households will temporarily save more and consume less than they would in the long run (as the Ramsey-Koopmans-Cass model requires). Similarly, firms will invest less because the cost of capital is higher than its long-run marginal product. Total output will be less than its natural level. How much less? That depends on the coefficient α, which measures the sensitivity of the economy's demand for output to this difference between the real interest rate and the natural rate.

One might try to estimate the value of α from data by econometric means, provided data are available for actual output, its natural rate of growth, and the real and natural interest rates, and also provided that the equation is approximately right about how these variables are related. But the best economics has been able to do is provide values that produce stylized facts about the whole economy. (See the appendix to this chapter for some details.)

At the end of the equation is ε_t, the variable that represents exogenous shocks to total demand. In the DSGE approach this variable usually represents factors that affect the economy's demand for output but are unpredictable from inside economic theory (whence, again, the S for *stochastic* in DSGE). It represents the macroeconomic effects of changes in government budget and tax policy, or more nebulous influences such as consumer confidence or investor optimism or pessimism, all working through rational choosers' expectations for the future. Such random variables can quickly shift total actual output away from its natural level.

Similar thought-experiment reasoning will generate the more complicated DSGE equations from the simple long-run equilibrium equations. For our purposes, the most interesting of these is how the macroeconomist reasons from the simple $\pi_t = E_t \pi_{t+1}$, to the more complicated $\pi_t = E_{t-1}\pi_t + \phi(Y - \bar{Y}_t) + v_t$, that is, from an expected rate of inflation next year exactly equal to its current rate to a state where expectations are wrong and output is not at the normal, long-term level.

Recall that $\pi_t = E_{t-1}\pi_t + \phi(Y - \bar{Y}_t) + v_t$ is the Phillips curve equation. There is no long-term money illusion. But in the short run, agents with rational but fallible expectations can mistake increases in nominal prices for what they sell—labor or commodities—and rising nominal costs for what they buy as marks of real-price increases that signal increased demand. When inflation, π_t, increases beyond expectations, rational agents increase production and output from the natural rate, $\bar{Y}_t$. The amount of the increase in production depends on the difference, $Y - \bar{Y}_t$, of price increases in the next year beyond this year's expected rate of increase. How much depends on the coefficient ϕ. But all this is in the short term.

The DSGE models' microfoundations enable economists to plot the inevitable progress of the economy back to long-term optimality, the natural rate of output, $\bar{Y}_t$. Start with a supply shock, v_t, or a demand shock, ε_t, or anything that moves the real interest rate away from the natural rate. Changes in the values of coefficients α and ϕ, governing the impact of discrepancies between $\bar{Y}_t$ and Y_t, will also have an impact. Seeing how and why an economy moves back from these conditions to the calm sea of general equilibrium is crucial to seeing why they are DSGE models. The whole point of such models is that no matter where stochastic shocks have taken the economy, at every moment the economy is dynamic. It always moves back in the direction of a general equilibrium, no matter how many supply shocks, demand shocks, and mistakes rational agents make.

A diagram common to many macroeconomic texts makes transparent the process back to long-run equilibrium dictated by the model. All macroeconomic changes are stops on a round trip from where the economy started, the long-run general equilibrium.

Recall what Keynes wrote immediately after his immortal quip that in the long run we are all dead: "Economists set themselves too easy, too useless a task if in tempestuous seasons they can only tell us that when the storm is long past the ocean is flat again."[11] Easy or not, this is the task macroeconomics sets itself. Figure 6.2 plots the cruise ship's course away from the safe harbor of general equilibrium and back again. Recall the diagram in chapter 5 that drove Friedman's prescient 1960s argument against Keynesian policies of increasing inflation to reduce unemployment. New classical macroeconomics relies on the same argument, generalized.

Figure 6.2 charts the short-run impact and the long-run disappearance of a deflationary, recession-producing change in money prices. Plotted against price level, the long-run aggregate supply (LRAS) curve is vertical. There is no money illusion in the long run. This long-run supply curve is constant (labeled Y). It's the natural level of output ($\bar{Y}$ in the equations we've been examining) unaffected by nominal values like the price level. The economy's long-term equilibrium level starts out where the short-term aggregate demand curve AD_1 intersects the vertical long-run supply curve at point E_1.

Short-Run vs. Long-Run Effects of a Negative Demand Shock

1. An initial negative demand shock...

2. ...reduces the aggregate price level and aggregate output and leads to higher unemployment in the short run

3. ...until an eventual fall in nominal wages in the long run increases short-run aggregate supply and moves the economy back to potential output.

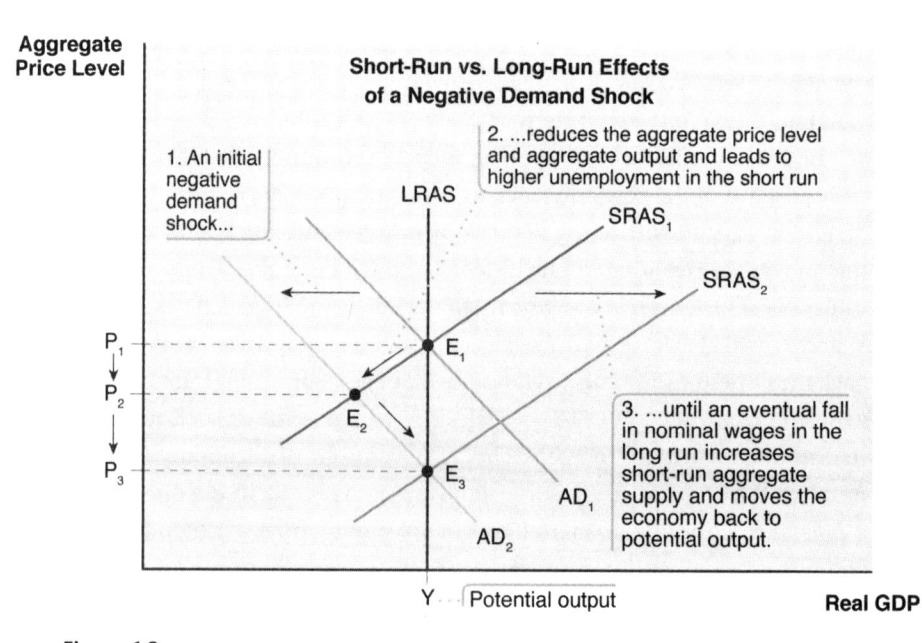

Figure 6.2

The curve is downward-sloping, as the first equation in our model requires: total output is sensitive to the money-price level because real interest rates depend on money rates and expected future inflation.

Suppose there's a recession—a reduction in demand at every price—resulting from some negative demand shock, ε. That moves the aggregate demand curve to the left, where it intersects at E_2 with a short-run aggregate supply curve that also goes through the original equilibrium point, E_1. This short-term curve of aggregate output, per the model, is sensitive to the price level, due to agents erroneously treating nominal price changes as signals of real-price changes. That's why it slants down and to the left: lower prices signal weaker demand. This moves the short-run equilibrium down and to the left, lowering output (real GDP) to E_2. But this fall in the nominal price level moves the short-run aggregate supply curve to the right, since labor costs, like all prices, have fallen, so more workers are hired. That results in the new equilibrium intersection of short-run supply and short-run demand, E_3. But that's where the economy started off: the same original level of output, Y, at a new, lower price level that is irrelevant to real and natural economic values. If we had begun with an inflationary supply shock, a similar diagram would plot the round trip back to the long-term general equilibrium, starting by moving the short-term aggregate supply curve to the right instead of the left.

The Ramsey-Koopmans-Cass microfoundations—households maximizing long-term utility and firms selling at marginal cost—guarantee that a long-term optimal equilibrium exists. The DSGE equations plot the path of an economy of fully rational, utility-maximizing households and revenue-maximizing firms back to this optimum whenever the economy is disturbed.

Finally, we can see that macroeconomics is just more microeconomics. Robert Barro, one of the leading new classical macroeconomists, introduces his textbook on the subject thus:

> The macroeconomic approach in this book is a continuation of the economic reasoning used to explain the behavior of individual households and businesses. Here we apply the same economic *science* to understand the workings of the whole economy. . . . A central theme of this book is that a more satisfactory macroeconomics emerges when it is linked to the underlying microeconomics. "More satisfactory" means, first that the macroeconomic theory avoids internal inconsistencies and second that it provides a better understanding of the real world.[12]

Outside observers of economic theory may be wondering whether we're trapped in a hall of mirrors, or maybe a revolving door. Recall from way back in chapter 2 how economists in the 1950s dealt with the falsity of their idealized model of RCT. Their rationale for using the RCT model was that it's a predictive device or a way of capturing market rationality, but there was no need to take it seriously. Indeed, that would be to mistake economics for psychology. Remember the founding philosophical document of postwar economic orthodoxy, Friedman's "Methodology of Positive Economics." Ever since 1953, economists have been fending off criticisms of RCT as misplaced and unwarranted by reinterpreting microeconomic theory, denying that the theory is about or tested by the behavior of individual consumers and firms. Back then Friedman famously rejected evidence of individual agents' thought processes as irrelevant to microeconomic theory. He suggested that asking economic agents about their choices was no more relevant to economics than asking octogenarians about their habits was useful to gerontology.

Then, famously, Friedman's student Gary Becker provided a powerful demonstration of the independence of microeconomic theory from any claims about individual consumers and firms in "Irrational Behavior and Economic Theory." Recall from chapter 2 that Becker's objective was to show that the main results of microeconomic theory would obtain regardless of whether actual individual agents were habitual—always attempting to conform their later behavior to their previous behavior—or random—choosing bundles of commodities by employing a chance device. No matter how agents choose, the result will still be downward-sloping demand curves. The theory of consumer behavior and the theory of the firm were, in Becker's words, really just disguised theories of market rationality.

Now, three generations of economists later, rational expectations macroeconomics demands we take the individual economic agent's thought process seriously. Economic theory had to start with firms' and consumers' observations of the economy and their inferences from those observations. We must tailor our macro theory to the fact that economic agents are fallible but continually update and correct their expectations over time.

Why Require Microfoundations?

Let's return to the comparison we made way back in chapter 2 between the RCT model as an idealized explanation of the laws of supply and demand and the billiard ball model as an idealized explanation of the gas law. It will show how different the economist's demand for microfoundations is from the apparently similar demand present in the natural sciences. It will perhaps finally make clear what is really going on in economic theory.

Recall from chapter 2 the story about how in the 1600s Boyle, Charles, and Gay-Lussac began running experiments and collecting data that suggested mathematical relations between pressure, temperature, and volume. By the nineteenth century these data were precise enough to support the ideal gas law, $PV = kT$, and give a value for the constant k. But further experimentation provided more data points for different gases at higher pressures and temperatures. The new data showed the ideal gas law was only approximately true, and only held for a certain range of values of P, V, and T. To explain the discrepancy, chemists began to hypothesize that there were unobserved "atoms" with different physical properties that made up different gases—the billiard ball model of the kinetic theory of gases. No one took these "atoms" seriously, however, until hypotheses about the differences among atoms and differences between the gases that they composed led to changes in the gas laws: changes that improved their predictive power for new data points. The accumulation of experimental data eventually enabled chemists and physicists to reduce the idealizations of the ideal gas law and convert it into an equation that explained and predicted far more data to far greater precision. This was the van der Waals equation:

$$[P + a(n/V)^2](V/n - b) = rT.$$

The microfoundations of the gas laws provided by the billiard ball model are evident in this equation and are responsible for its predictive accuracy. In the van der Waals equation, a measures the intermolecular forces that the ideal gas law idealizes away and b measures the radius of the gas molecules, which the ideal gas law also sets at zero. At first these microfoundations, the values of a and b, couldn't be measured directly. Their values could only be estimated by working backward from experimental data. But once their values were estimated, more predictions could be made about the behavior of gases. So successful were these predictions that by the end of the nineteenth century physicists and chemists began to accept the reality of the billiard ball model's atoms. The microfoundations of the behavior of gases provided by molecules earned their explanatory status by enhancing the prediction of experimental data.

Why demand microfoundations? In physical science it's the way to improve the predictive power of macro models and to explain why the predictions are better. We could have told the same story about advances in genetics from Mendel to Watson and Crick or advances in neuroscience from Pavlov to Kandel.

It is safe to say that this recursive sequence—from original macro data and macro model through microfoundations to increasingly accurately predicted macro data to revised macro model to a still better micro model—is not exemplified in the development of DSGE models. These models didn't emerge from data in the first place, and their microfoundations didn't result from attempts to explain data regularities, even ones characterized as stylized facts.

At the outset of macroeconomics in the 1940s, the development of Keynesian models seemed to parallel that of macro models in genetics or chemistry. Econometricians sought to model data regularities among macroeconomic variables—national income, employment levels, the money supply, the interest rate, and more. The resulting models ranged from a simple Keynesian three-equation model to one for the United States consisting of about 150 equations.[13] Keynesians didn't seek microfoundations for their macro models because they never had enough confidence that the models were sufficiently accurate over a long enough time period. By the 1970s the failure of Keynesian models to predict economic developments and guide policy led some economists to despair of the approach. Recall Lucas and Sargent's hand-wringing about the sorry state of Keynesian macroeconomics: "That these predictions were wildly incorrect and that the doctrine on which they were based is fundamentally flawed are now simple matters of fact, involving no novelties in economic theory."[14]

DSGE models emerged from the wreckage. But they didn't emerge from data the way macro models in natural science typically arise from data. They came straight from microfoundations. In the words of an important macroeconomist, "Modern DSGE models are based on microeconomic foundations."[15] And where did the microfoundations come from? From classical microeconomics, inspired by Smith's invisible hand, via RCT and the first theorem of welfare economics. Unlike microfoundations in natural science, these microfoundations didn't earn their explanatory place by increasing predictive power.

By imposing on themselves what Lucas and Sargent called "the discipline" of optimizing agents and market-clearing models, new classical macroeconomists reproduced Keynes's flat ocean after the storm. They did so by deduction, imagining themselves as rational agents and deducing how they would respond to economic changes—real changes and their money-priced symptoms.

If the provision of microfoundations had been driven by the imperatives driving micro-modeling in natural science, the DSGE models derived from them would have done better at modeling macro data than the Keynesian models they superseded. But by and large, this is not what happened. DSGE models have done no better than their 1960s predecessors. Some influential economists have complained that they are worse. Robert Solow, awarded the Nobel Prize for his work in macroeconomic growth theory, has written: "We want macroeconomics to account for the occasional aggregative pathologies that beset modern capitalist economies, like recessions, intervals of stagnation, inflation, 'stagflation.' A model that rules out pathologies by definition is unlikely to help. It is always possible to claim that these 'pathologies' are delusions, and that the economy is merely adjusting optimally to some exogenous shock. But why should reasonable people accept this?"[16]

New Classical Macro and the Subprime Mortgage Recession

Few events in the history of new classical macroeconomics illustrate its method of theory formulation and modification more clearly than the period after the subprime mortgage meltdown of 2007–2009. Most economists and many others will recall the premature boast Lucas advanced toward the end of what economists called "the long moderation" from 1980 to 2007: "Macroeconomics in this original sense has succeeded. The central problem of depression prevention has been solved, and has in fact been solved, for all practical purposes, for many decades."[17] Then came the financial sector's meltdown, which precipitated an economy-wide departure from anything like market-clearing equilibrium.

Since the crisis DSGE macroeconomists have accepted that the models developed before the subprime mortgage meltdown require revision. The revision involves adding "frictions" to the models, frictions of a sort that macroeconomists were already familiar with, that explain why economies do not invariably and quickly move away from shocks toward the long-term natural level of output that DSGE models require.

Even before the financial crisis of 2008, DSGE modelers often hypothesized barriers to individual firms flexibly adjusting their selling prices in light of changes in the cost of inputs and demand for their products. The resulting "sticky prices" were frictions that slow down the round trip from general equilibrium to a demand or supply shock and back to general equilibrium.

DSGE theorists who claimed prices are sticky labeled themselves *New Keynesians*. The label they gave themselves was seriously misleading to the outsider. Old man Keynes's main argument was that the economy rarely, if ever, achieves or even approximates a market-clearing equilibrium and isn't always naturally moving toward it. But the New Keynesians accepted the new classical assumption that the economy does move toward market-clearing equilibrium. They rejected Keynes's key claim and sought to identify short-term impediments the economy would have to overcome in order to regain general equilibrium. Additionally, Keynes and his followers were prepared to grant money prices a real role in the economy and even sanction the existence of a money illusion that governments could exploit. The New Keynesians submitted to RCT's market-clearing "discipline" as imposed on macroeconomics by Lucas and Sargent. The New Keynesians focused on the labor market and tried to find reasons driven by RCT that firms and workers would refuse to lower wage rates to market-clearing levels for as long as they could. Instead of cutting wages during a recession, New Keynesians argued, rational firms will instead reduce their labor force while paying remaining employees an above-market-clearing wage to retain their best workers. That makes the price of labor sticky, too high to clear the market of all those willing to work, thus creating involuntary unemployment. Without irony, they called these prices *efficiency wages*.

Introducing these frictions to the labor market did not enhance the predictive power of DSGE models. But it did give economists reasons not to expect a rapid return to equilibrium after an exogenous shock.

Then came 2008, after which frictions in financial markets had to be added to frictions in the labor market. DSGE modelers identify two kinds of financial frictions that were ignored prior to the subprime mortgage crisis and that could be incorporated into models to help render the crisis consistent with DSGE models. One kind of friction results in transactions among financial institutions that generate *roll-over crises*. Small but sudden drops in asset prices (e.g., of houses) can force banks to sell off suddenly cheaper assets—to roll them over. Lower asset values mean larger liabilities. If this happens to many banks, and the only prices buyers are willing to pay will not cover banks' liabilities, no one will extend credit to banks and they won't be able to purchase one another's assets at any price, whence a rollover crisis. The result is a financial shock that moves the economy away from equilibrium in its most important market—the financial market. That market catastrophe then moves the rest of the economy away from equilibrium as well.

A second friction emerges from a sudden exogenous increase in the risks to all businesses that borrow. A deep recession produces such an increase. DSGE models tell us that recessions end because the reduced level of economic activity decreases interest rates and so increases investment opportunities. But there is an increased risk friction operating in the opposite direction, "reflecting the view of lenders that firms represent a riskier bet" at any interest rate.[18] These frictions and the phenomena of sticky prices and sticky wages from the New Keynesians were built into DSGE models after 2008. They were added as after-the-fact explanations of stylized facts about the subprime meltdown contrived by thinking through how rational agents would respond to exogenous factors and money-price changes that increasingly diverged from the long-term real prices that everyone would eventually figure out.

More than one distinguished Nobel laureate had by this point lost confidence in DSGE models and begun to be perplexed by the continued attachment to them that most macroeconomists manifested. Joseph Stiglitz wrote:

> To be sure, as many users of DSGE models have become aware of one or more of the weaknesses of these models, they have "broadened" the model, typically in an *ad hoc* manner. . . . There has ensued a Ptolemaic attempt to incorporate some feature or other that seemed important that had previously been left out of the model. The result is that the models lose whatever elegance they might have had. . . . And with so many parameters, macro-econometrics becomes little more than an exercise in curve fitting. . . . The DSGE models fail in explaining these downturns, including the source of the perturbations in the economy which give rise to them: why shock[s], which the system (in these models) should have been able to absorb, get amplified with such serious consequences; and why they persist, i.e., why the economy does not quickly return to full employment, as one would expect to occur in an *equilibrium* model. These are not minor failings but go to the root of the deficiencies in the model.[19]

Stiglitz found the fault in the microfoundations that new classical macroeconomists have so ardently insisted on.

DSGE models seem to take it as a religious tenet that consumption should be explained by a model of a representative agent maximizing his utility over an infinite lifetime without borrowing constraints. Doing so is called *microfounding* the model.[20]

The microeconomics of the basic competitive model—as formulated in Arrow and Debreu [1954, the proof of the first theorem]—has been shown to be flawed by 40 years of economic research. Why should we expect a macroeconomic model based on such microfoundations to work? ... Macroeconomics is supposed to provide us with models of how the economy *actually* behaves, rather than how it might behave in a mythical world of infinitely selfish people but among whom contracts are always honored.[21]

Macroeconomics as Moral Philosophy

What if Stiglitz is wrong about what macroeconomics is supposed to provide? Maybe it models not how the economy *actually* behaves, but rather how it *should* behave. In that case many of the differences between the DSGE model and macro models in empirical science begin to make sense.

Instead of hypothesizing that market economies always tend toward natural, optimal levels of output, one might hypothesize that economies won't do so by themselves. Then one might seek to design a set of institutions that will drive a market in the direction of optimality. DSGE models could be treated not as descriptive, or at least not entirely descriptive, but as *prescriptive*, in a certain measure. They reflect not simply a claim about the character and operation of the macroeconomy but also a set of hypothetical imperatives about how the macroeconomy could operate and how it should be managed in light of what a society aims to achieve. The DSGE model would in significant measure turn out to be an account of how certain institutions should be designed and operated to attain the desired outcome: the market-clearing, welfare-optimum general equilibrium that Kenneth Arrow and Gérard Debreu identified in the first theorem of welfare economics.

The fact that the joint behavior of consumers and firms, if completely rational, provably results in a preference-maximizing general equilibrium may be a reason to (re)organize a real economy to allow it to do so and to take steps to encourage a real economy to move in that direction. We don't have to treat the search for frictions that an economist like Stiglitz stigmatizes as ad hoc devices to protect the theory from empirical disconfirmation. We can treat it as part of the diagnosis of institutional obstacles to the real economy's attainment of the optimal general equilibrium. Sticky prices and wages need to be eliminated due to their impact on efficiently attaining the natural rate of output. Financial frictions are to be mitigated by improving the design of financial institutions to more accurately calibrate and trade risks.

The incorporation of frictions in DSGE models had immediate policy implications in developed economies, especially the United States. Features of the Dodd-Frank Wall Street Reform and Consumer Protection Act of 2010 are a primary example. Long before 2008 wage rigidities and sticky prices had been targeted by deregulation and by erecting obstacles to labor unions (so-called right to work laws). The Fed's policy of maintaining price stability reduces firms' need to change prices and so reduces the drag that sticky prices can impose on the economy's approach to its natural level of output.

To see the role of institution design and policy prescription in macroeconomic models, let's go back to the monetary policy equation, the last of the five equations in our DSGE model. It's the formula for the nominal rate of interest:

$i_t = \pi_t + \rho + \theta_\pi (\pi_t - \pi_t^*) + \theta_y (Y_t - \bar{Y}_t)$.

Translating this equation into words is not easy and then holding it in mind takes a good working memory. The formula equalizes the nominal rate of interest in year t, i_t, to that year's actual rate of inflation, π_t, plus the natural rate of interest, ρ, plus a multiple of the difference between actual inflation in year t and the central bank's target for inflation in year t, θ_π $(\pi_t - \pi_t^*)$, plus a multiple of the difference between actual output in year t and the natural rate of output. The equation's use of the target for inflation, π_t^*, makes explicit reference to an institution: the central bank. It makes implicit reference to others, including the commercial banks that borrow from, purchase from, and sell to the central bank. In the model there are two parameters, θ_π and θ_y, that the central bank estimates. The central bank controls the nominal interest rate and raises or lowers it by amounts determined by its guesses regarding the values of θ_π and θ_y. Macroeconomists usually call this equation the Taylor rule, after the economist John B. Taylor, who suggested it. The Taylor rule requires the central bank to keep the real interest rate at or above the nominal interest rate. Doing so prevents an economy governed by a DSGE model from spiraling away from its long-run natural level of output to lower, less optimal levels.

Long ago Adam Smith argued that the competitive economy's invisible hand made everyone better off. As we saw in chapter 2 it took about 175 years for mathematical economists to form this conjecture into a theorem—Arrow and Debreu's first theorem of welfare economics, which establishes the existence of a general equilibrium and its Pareto optimality in satisfying preferences. The theorem and the conditions under which it holds provide a set of implicit prescriptions about how to design institutions that will secure the optimum whose existence they establish. The mathematics thus drives the politics of institution design—the classical and new classical economists' demand for free markets. Even if economies never attain such an optimal general equilibrium by themselves, the proof that general equilibrium is attainable under certain conditions is supposed to be reason enough to attempt to realize these conditions in the real world, to rearrange reality to be more like the kind of economy the proof requires.

Macroeconomic theory's demand for microfoundations is thus not anything like the demand for microfoundations in empirical science. It isn't driven by an imperative to increase the range and precision of the macro model's predictions and explanations but is motivated by the discipline's commitment to the welfare-optimizing character of behavior in accordance with the micro theory. Grounding macro models, like the DSGE ones, on micro models makes the attainment of the optimum equilibrium a macroeconomic possibility. So understood, when the DSGE models and the data they purport to explain and predict diverge, economists have some scope not to blame the model. Instead they can blame reality and advocate for the design or redesign of political and economic institutions to bring reality into better accord with the model. The commitment to the economy having a long-run natural level of output is not entirely, or perhaps even largely, a matter of empirical evidence. It is a political, moral, normative objective that economists suppose they share with many other rational agents.

As we'll see in chapter 7, the irony of this outcome is that in order to fix the economy, new classical economic theory has to give way to quite a different tool.

Appendix: DSGE Modeling of the Data

Here is an example of how DSGE models are confronted with macroeconomic data from the article "Understanding the Great Recession" by Lawrence Christiano, Martin S. Eichenbaum, and Mathias Trabandt. The paper does not address the causes of the subprime mortgage crisis that precipitated the recession. It couldn't. Before the crisis DSGE models were silent on financial market frictions that were added afterward to make the model compatible with the collapse of the financial markets.

The model the authors advance seeks to identify which of the exogenous stochastic shocks to the macroeconomy was the most important cause of the depth and duration of the subsequent Great Recession. They identify eleven macroeconomic variables as indicators of the economic decline precipitated by the subprime mortgage crisis in the United States: GDP, inflation, the interest rate the Federal Reserve charges banks on loans, the unemployment rate, the employment rate, the size of the labor force, aggregate investment, consumption and wage rates, the vacancy rate for the economy, and the average rate at which new jobs are found by job seekers. Christiano, Eichenbaum, and Trabandt plot the range of data and compute the mean of the values. Then the values of the coefficients and variables of their DSGE model are adjusted, post hoc, in ways that are mutually consistent and intuitively reasonable to the modeler. Plugged into the equations of the model, the theory produces *impulse response curves*—how the model's variables respond to one another over time after an initial shock. These curves are compared with the mean values of the data and then adjusted to produce a best fit.

In figure 6.3, the line that extends beyond the gray ("min-max range") in each of the data estimates gives the values for each of the eleven endogenous variables that the model has been calibrated to generate by setting the values of its parameters and coefficients. The fit of the line given by the model for any one of these eleven variables is constrained by the need to adjust the values of these parameters and coefficients to most closely fit the line for the other ten values.

Inspection shows that some of the model's values can be fitted closely to the mean of the data, for example the labor force, while others are outside even the range of the actual data estimates, such as the job finding rate and the wage rate.

Impulse curves generated by plugging numbers into a model are a stock-in-trade of macroeconomics. When central banks combine such DSGE models to project future economic outcomes, the results have never been well confirmed even over near-term time horizons. Instead new data have

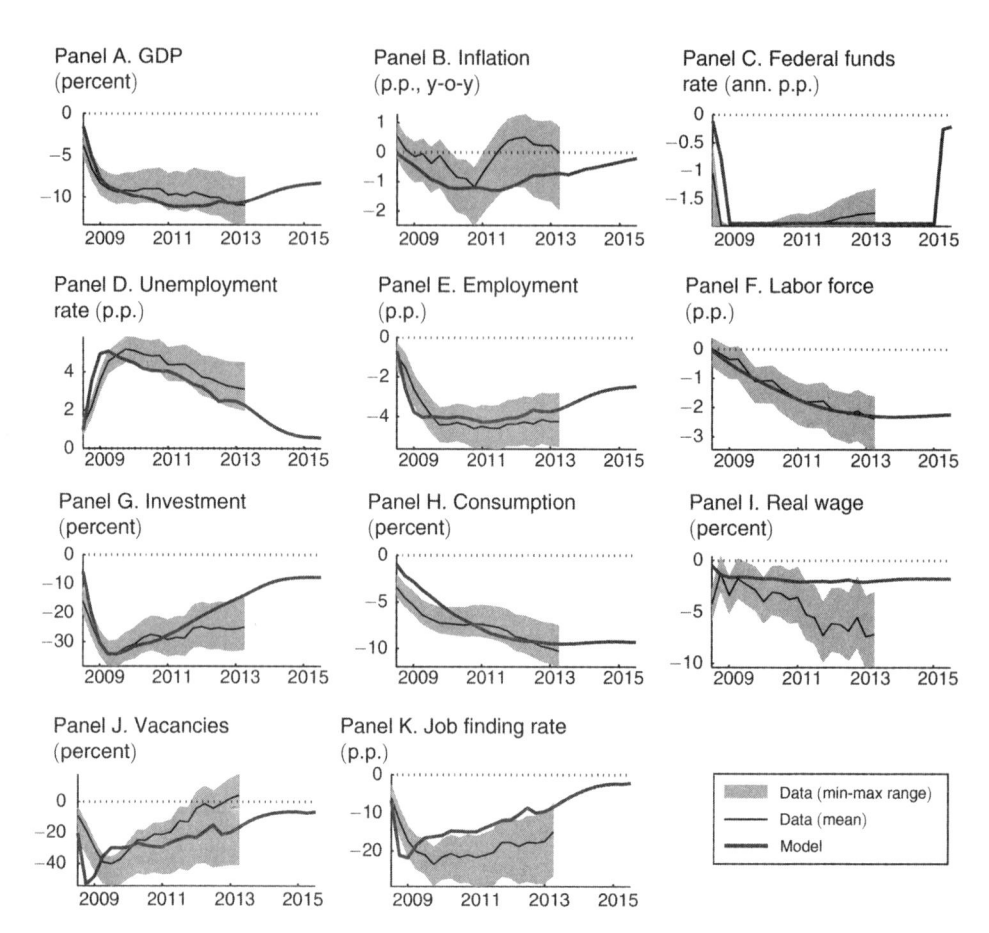

Panel A. GDP (percent)
Panel B. Inflation (p.p., y-o-y)
Panel C. Federal funds rate (ann. p.p.)
Panel D. Unemployment rate (p.p.)
Panel E. Employment (p.p.)
Panel F. Labor force (p.p.)
Panel G. Investment (percent)
Panel H. Consumption (percent)
Panel I. Real wage (percent)
Panel J. Vacancies (percent)
Panel K. Job finding rate (p.p.)

Data (min-max range)
Data (mean)
Model

Figure 6.3
From Lawrence Christiano, Martin S. Eichenbaum, and Mathias Trabandt, "Understanding the Great Recession," *American Economic Journal: Macroeconomics* 7, no. 1 (2015): 152.

been used to update values of the parameters and variables in DSGE models. The models themselves can't be tinkered with except in ways that reflect the operation of their RCT-derived microfoundations.

If macroeconomics worked the way modeling does in other domains of empirical science, it would have started with data, then framed mathematical equations that relate the mean values of the data sets. Once tested against new data, the equations would have been revised. Only then, if the equations showed some reliability in predicting new data, would a search for microfoundations—an underlying causal mechanism—be sought. Such a micro model would itself be accepted, at least tentatively, if and only if it led to further improvements in the macro models' accommodation to older data and, even more importantly, its ability to predict still newer data.

Instead the DSGE models begin with microfoundations and then economists explore how adding frictions make the models consistent with data that had disconfirmed them. What new classical macroeconomics won't do is give up the "discipline" of optimizing agents and market clearing that guarantees general equilibrium.

7 Economic Theory's War against Profit

From here on in we'll deal with the problem that both economic theorists and we outsiders to the theory face: market failure and how to respond to it. Economic theorists have to worry about the problem because it's pervasive in the real economy. We non-economists have to face it because we are its victims—at least we are if we're working stiffs earning a living and investing our savings for retirement. If it was curiosity that drove your reading until now, well, self-interest might be enough to keep you reading.

Market failure is a technical but simple concept once you grasp the first theorem of welfare economics. It's when the free market—the uncoerced exchange of goods between economic agents—fails to be allocatively efficient: when it undersupplies what buyers really want and can afford, or oversupplies what buyers are less interested in purchasing. Market failure happens when there are shortages or surpluses even though everyone is behaving as RCT requires. The first theorem proves that this can't happen in a perfectly competitive economy. When it does happen in any real or modeled economy, theory tells us something has gone wrong and needs to be fixed.

The next two chapters give us the tools to identify market failures and chart their effects on us. They will be a mix of more theory and our own experiences as sellers of labor and buyers of everything else.

The great thing about the perfectly competitive economy is that there is no way to game the system. There is nothing anyone can do to get more than they actually earn, be paid more for their product than it's worth to everyone else, or buy anything at a price below its cost. That's because in the perfectly competitive market everyone is a price taker. No one can influence the price of anything by trying to undercut the market or corner it. In a perfectly competitive market there are indefinitely many buyers and sellers, all trading the same commodity. Everyone knows everything they need to know about the commodity and how to make it, and about the other traders. If one seller withholds goods to raise prices, there are enough other sellers that it won't work. Try selling at a price higher than the market sets and no one will buy from you. Sell at a lower price and your stock will sell out so quickly there won't be a noticeable change in the market price, and you'll lose money in the bargain. It comes to this: In a perfectly competitive market, every buyer and every seller can just ignore the choices other people make. All they need to know is the market price. No one is offering a better deal—and no one can, if they want to stay in business.

As the proof of the first theorem promises, the perfectly competitive market is *allocatively efficient*: it delivers all the available inputs to productive processes that will result in the largest possible bundle of commodities consumers actually want and then it distributes them Pareto optimally. It does this by making everybody a price taker: they get no more for what they sell than what it's worth to the whole market, and they can't buy anything at a price that doesn't fully reflect the costs of its production. The perfectly competitive economy wastes nothing; every input is set to optimal productive use, and none of the resulting output is left unsold.

This is Smith's invisible hand at work, rewarding everyone with enough to make them do their best for the rest of us but no more than it takes to secure this outcome. Nothing that can be made and sold is wasted or wanted. And it's all done by the price mechanism.

Goodbye to Profit, and Good Riddance

Recall the irony that there is no room for money in the competitive economy of RCT. Well, there is no room for profit, either. Karl Marx famously argued that capitalist competition must produce the immiseration of the proletariat by forever driving down the wage rate of labor to subsistence levels. Ironically, it turns out that the perfectly competitive market drives down the capitalist's returns to subsistence levels as well!

The conditions that turn Smith's conjecture into the proof of market optimality exclude the very thing we think of as motivating the butcher and the baker and everyone else in the economy. Economic theory forces us to completely rethink our idea of profit, in fact to give it a bad name and seek to drive it from the real economy, in the name of making the market more efficient and closer to perfectly competitive.

In the perfectly competitive economy, profit is impossible; it doesn't exist; there is none! Just try to sell something for a dollar more than it cost you to make it. With an indefinitely large number of buyers and sellers in the market, only two things can happen: no one will buy from you at all, or if someone is foolish enough to pay you more than what everyone else is charging, there will soon be another seller offering the same product to these knuckleheads at cost plus ninety-nine cents. You respond by lowering your price to cost plus ninety-eight cents, and the next thing you know, you're both back to selling at the original market price. In perfect competition all you ever get for what you sell is your actual cost of production, including of course your own labor costs, the cost in wear and tear to your tools, and the rent for your workspace. The perfectly competitive market squeezes out all profit in its drive for efficiency in the allocation of inputs. Everybody, including firm owners, is paid for their input, but all they get is the minimum needed to convince them to contribute.

It's the six assumptions in chapter 3, from which the first theorem of welfare economics follows, that make profit impossible. The allocatively efficient market will have a large number of buyers and sellers, all of whom will be rational. Those are the first two assumptions. If one seller tries to jack up prices, buyers can turn to many other sellers. If one buyer tries to depress prices by not buying, there are too many other buyers for their actions to affect the market price.

In perfect competition, everyone has to know everything about everyone else's production and consumption choices—what they actually choose to buy and sell at what prices. The presence of complete futures markets enables them all to plan rationally for the future, including by producing long-term capital goods, buying insurance, and locking in the commodity supplies they'll need later by paying for them now. If everyone is rational and knows everything relevant to exchange, no one can exploit anyone's ignorance to make an extra buck.

Then there's the requirement of constant returns to scale in production: if you double all inputs, you will exactly double outputs. There can be no less (i.e., diminishing returns to scale) and no more (increasing returns to scale). This means that no matter how big a business becomes, it can't lower the cost of each output unit it sells and thus can't make higher profits by being bigger. In fact, it can't make any real profits at all. The only way to do so would be to reduce its production costs below those of other firms. But if everyone is using equally productive production processes, that is also impossible. Economic theorists realized that they needed to assume constant returns to scale if they were going to be able to provide for the existence of a Pareto optimal general equilibrium in an economy that includes producers as well as consumers of goods.

If the perfectly competitive economy rules out profit, why bother making any effort at all? Well, perfect competition rewards everyone for their inputs to production; it just doesn't allow anyone to rip off others—to achieve returns greater than needed to elicit their contribution to output, which would be a waste of resources that could be used elsewhere to produce more at lower cost or buy more to satisfy preferences more fully.

Economists have a word for these excess profits: *rents*. The word originates from the notion that when land is rented, the owner doesn't do useful work to get the rent. The "rent" every price setter makes is an unearned return that could have been put to better use by the buyers it was taken from. Rents can't arise in any economy that is optimally efficient, that produces the maximum amount of the things people really want from all the available production inputs. Rents destroy the benefits of perfect competition. There is no more room for profit in the perfectly competitive economy than there is for money.

Why should we take seriously a model of the economy that makes economic profit impossible? Economic theorists insist that perfect competition *does* allow for real economic profit, or rents. They do so by distinguishing the long run, in which there are no real profits, from the short run, in which there often are. In the short run a firm can introduce a new production process that lowers its costs or a new commodity no one else is selling and many consumers want. In the first case it can scoop up the difference between its new, lower costs and the market price that reflects everyone else's higher costs of production. In the second case it can charge any amount for the new product, regardless of the cost of production; it can be a price setter and make economic profits that way. Eventually, in the long run, the rest of the market will catch up: other producers will adopt the new production method or start selling the same new product (or a similar enough substitute) and the short-term rents will be whittled away to nothing. So long as firms can enter the market or copy the latest improvements in production and products, the market will return to the zero-profit Pareto equilibrium.

Entrepreneurs, take heed! Make hay while the sun shines. The reason you're in business is to make rents by doing what no one else can do, meeting a need consumers didn't even realize they had, or finding an entirely new way of meeting an existing need. In the perfectly competitive market there's no room for you in the long run. In its war on profit, economic theory makes a casualty of entrepreneurship. No wonder no textbook of the theory mentions it, still less helps us understand it.

We've seen the long run–short run distinction before, especially in chapters 5 and 6. There we noted that economic theory calibrates time periods self-referentially: the long term is the period an economy takes to get to equilibrium. How long is the short term? Well, it lasts as long as it takes for profits to completely disappear, to be beaten out of the economy by competition. How long is that? It depends on how far the market is from the six conditions that guarantee the truth of the first theorem of welfare economics—the Pareto optimality of market-clearing general equilibrium.

Unlike in the perfectly competitive market, in real markets there's never a time when someone isn't making rents, producing inefficiencies while enriching themselves. But unlike in the real market, it's hard to see how short-term opportunities for real economic profits could ever emerge in the perfectly competitive market. Think about it: under perfect competition, everyone is working as hard as they can, making stuff at zero profit and consuming leisure and commodities. Nobody has anything extra lying around unexploited that could be used for tinkering, inventing, discovering, or taking risks to find better ways of doing things. Entrepreneurship always requires some start-up resources—some investable capital not already dedicated to earning zero-profit returns. What entrepreneurship needs is what only successful rent seeking can provide—real economic profits. It looks like a bit of sleight of hand when economic theory helps itself to this resource just to allow for real profits as short-term possibilities to be quickly eaten up by perfect competition. Unless rents keep mysteriously entering the economy from the outside, via exogenous forces, short-term economic profits are no more possible than long-term real profits in the model of perfect competition.

When economists allow for real economic profits, but only in the short term, they're just admitting that the model doesn't have room for profit and we shouldn't expect economic theory to tell us more about it or the entrepreneurs who realize it any more than economic theory can tell us about money. It's not just *Hamlet* without the prince—it's *Hamlet* without Ophelia or Horatio, either.

Insofar as economists and others extol the virtues of competition and strive to unlock the magic of markets to fulfill the mission of the invisible hand accorded them by Adam Smith, they must oppose profit as their enemy. That, at any rate, is what the theory teaches.

Economic theory has a name for markets in which profits are made. They are places of *market failure*. The existence of profits is always due to the presence of traders with *market power*. Besides their technical meanings, these terms are labels for things that economic theory disapproves. *Market power* is exercised by any buyer or seller who can set prices, who isn't a price taker. The exercise of market power produces *market failure*: the failure of a market to be allocatively efficient and provide the largest bundle of goods and services people actually want and can pay for given available inputs.

The Harm That Profits Do: Monopoly

To get a feel for how baleful the impacts of profits are, we need to pursue economic theory's analysis of two clear forms of market failure: monopoly, where one seller has complete market power, and monopsony, where the market power is in the hands of a single buyer. Most people are more familiar with monopoly, but we should be much more concerned with monopsony. Both cause real-world trouble, but the problems they raise for economic theory are so radical that a whole new approach is required to understand them—one we'll spend the rest of this book addressing.

When it comes to monopoly, every textbook in economic theory shows the same graphs. They are the clearest illustrations of what's so bad about real economic profits—rents—for almost everyone but the monopolist. Economists have been using these graphs to illustrate the way monopoly works for more than a century. Figure 7.1 displays the standard graph for monopoly. It's the model of another stylized fact economists seek to explain by deriving it from RCT. The diagram contains a lot of information, along with some "dis-" or "mis-information" about the model of monopoly.

The first thing to notice is that in the middle of the graph, where the demand curve crosses the marginal cost curve, we can find the price and quantity for the commodity traded at perfect competition, the ones that price takers would have to accept. The marginal cost curve is the supply

Economic Theory's War against Profit

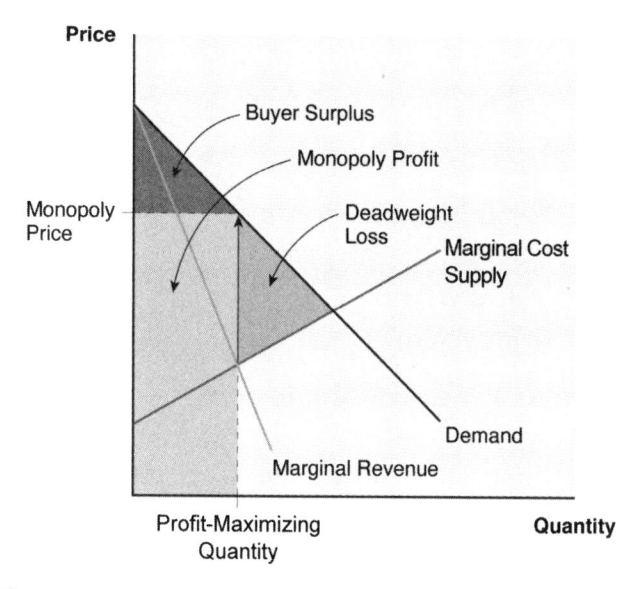

Figure 7.1
The monopoly model.

curve for firms in this market. Why? In perfect competition, where there is no profit, the marginal cost of production—the cost of building the last item sold—is equal to the price that the market sets for that item. The firm will produce and sell every unit it can up to this point and none beyond it. The firm would lose money for every unit it sold at a price below what it cost to make that unit.

The second thing to note is the triangle labeled "buyer's surplus." This is where the model begins to represent things that are important in economic theory's diagnosis of the problem with profit. To see what's wrong with it, let's look at the graph for perfect competition in figure 7.2.

Let's imagine this is a market for apples. The total quantity of apples traded is Q1, and the price at which each apple is sold is P1. How much would a consumer be willing to spend on the very first apple that is made available for sale, all the way to the left on the x-axis?

If the consumer is like us, they would pay a lot for it—much more than they would pay for the apple that would bring the total quantity of apples consumed from zero to Q1 on the graph. This is what the principle of diminishing marginal utility says. Economic theorists can't actually invoke this principle because they deny themselves any appeal to psychology. (The

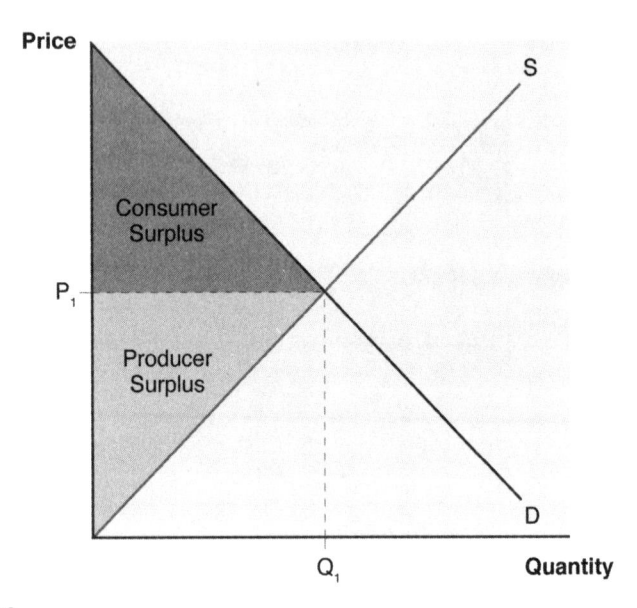

Figure 7.2
The consumer's surplus in perfect competition.

closest they can come is asserting a principle of diminishing marginal rates of substitution between, say, apples and oranges. The fewer apples a consumer has already had, the more oranges they'll give up in exchange for an apple. Why? Don't ask; that's psychology.)

The consumer would pay more for the very first apple consumed than for the second, more for the second than for the third, and so on all the way down the demand curve to where the consumer pays the price P1 for the apples bought at Q1. But the consumer only has to pay P1 for every apple—the most wanted first one, and the slightly less wanted second one, through the last one they're willing to buy at P1. All the utility that consumers get from the total number of apples they consume is the triangle above and to the left of the price and quantity sold under perfect competition. The first and last apples purchased cost the same, but the utility derived from eating the earlier apples is much greater than the utility derived from eating the last of them. The consumer doesn't have to pay for the extra utility derived from the first, second, or third apple. Under perfect competition, consumers pay for each increment of utility only what they would pay for the minimal amount of utility derived from the last apple purchased.

Economic Theory's War against Profit

Three cheers for perfect competition! It delivers some quantity of "consumer's surplus" of utility—or would, if there were utilities that came in units that could be added up and compared. But since there are no such units, the consumer's surplus isn't a measurable size. Even if it were, even if utility existed in addable amounts, we wouldn't be able to measure it unless we could measure demand curves. But we can't do that either because their shapes and slopes change too rapidly. Of course in perfect competition, where everyone is a price taker, no one needs to measure demand curves—or supply curves, for that matter. All they need to know is the price fixed by the two curves together.

The consumer's surplus isn't something economists can measure. It's another stylized fact. Never mind that, the economist tells us; the graph is just a model. It doesn't have to be realistic to be useful.

Now we can see what's so bad about monopoly: it destroys some of the consumer's surplus. (We can't tell how much exactly, since we can't measure it, but never mind.) Even worse, monopoly doesn't pass along that utility to the seller or anyone else; it's just destroyed. Look back at figure 7.1. Instead of taking the price imposed by the competitive market, monopolists can impose any price they like. How high a price should that be? Once we add a marginal revenue line to the graph, we can answer that question. The marginal revenue is the amount the seller gets from the last unit sold. If they sell just one unit, the seller can charge a high price, since the buyer gets a lot of utility from the first unit consumed. But as the seller sells more units, the utility that consumers derive decreases (presuming diminishing marginal utility), so they'll pay less for each additional unit and the marginal revenue will fall. The marginal revenue curve falls faster than the demand curve. Meanwhile, the marginal cost of production is rising as the firm's demand for inputs to production increase. Recall that the marginal cost curve the firm faces is its supply curve.

Where the marginal revenue curve and the marginal cost curve intersect, the cost of making a unit to be sold and the revenue from selling that unit are equal. Above that level of production the firm doesn't want to produce or sell more, since it will make less in sales than it loses in costs to make the unit. Now trace a line straight up from the intersection of the marginal cost and marginal revenue curves to the demand curve and move left to the price axis. That will be the amount the monopolist should charge to get the biggest return. It will be higher than the price the firm can charge when it is a price taker under perfect competition. The total amount of revenue the monopolist makes is the rectangle that extends to the higher price on the demand curve. The extra revenue they make beyond what perfect competition would allow is the trapezoid above the marginal cost curve. That is economic profit: revenue the seller wouldn't earn under perfect competition, when the price they can charge is lower. It's money that consumers could and would spend on other stuff under perfect competition, but can't if they buy at the higher price the monopolist decides to charge.

This revenue has no economic justification. The monopolist makes rent at the expense of buyers, but in doing so produces inefficiency for the whole economy—the consumer loses utility that the monopolist doesn't capture. This is the amount labeled "deadweight loss" in figure 7.1. This is a further loss to be figured in along with the reduced consumer surplus. Recall the smaller buyer's surplus triangle in figure 7.1 compared to the consumer's surplus in figure 7.2. The deadweight loss triangle represents the utility consumers would get under perfect competition that no one attains when the monopolist reduces the buyer's surplus.

The monopolist's profit (the gray trapezoid in figure 7.1) is money taken from consumers, who pay a higher price for a smaller quantity of a good than could be purchased under perfect competition. This profit doesn't exist under perfect competition, when a seller's income is exactly equal to the cost of production.

Now we can see why monopoly reduces the efficiency of the economy, preventing it from producing the biggest bundle of commodities people want; why it reduces preference satisfaction below the level that perfect competition would provide; why it produces the wrong prices and the wrong quantities exchanged; why there is market failure. But this model of monopoly is more limited even than other idealized models are. Its limitations weaken its implications for analysis and for policy.

The model can't tell us how large the consumer surplus is under perfect competition or how much of it is expropriated by monopolists when they charge more than the perfectly competitive price and sell less than the perfectly competitive amount. In addition, the model cannot tell us the magnitude of the deadweight loss to the whole economy that results from the monopolist's pricing and production decisions. The triangles look large on the graphs but that's just an artifact of how the lines are drawn. For all we know the expropriated consumer surplus and the deadweight loss could be slight or vast.

As we've had occasion to regret before, the problem is that we can't get beyond models of stylized facts. The existence of consumer surpluses and deadweight losses are stylized facts, not ones quantitative amounts can be attached to. To do that we'd need to draw real, reliable, relatively constant demand, supply, and marginal revenue curves, connecting data points representing repeated behavioral choices that can be compared. In addition, we'd need units in which to measure consumer surplus and deadweight loss—not units of utility, which doesn't come in measurable amounts.

There is another problem obscured by the graph, a vast issue hidden by this model of monopoly pricing. It's one that compelled economists to seek an entirely new theory to explain economic behavior, which I'll spend the rest of this book introducing and applying: *game theory*, the study of strategic interactions.

The problem made invisible in figure 7.1 is one facing the monopolistic seller and the buyer. It doesn't exist in perfect competition, in which buyers and sellers are price takers who can buy and sell only at the market price. They don't know the shape of the demand curves, but they don't need to know them in order to figure out how much to charge and how much to pay. But in monopoly, buyers and sellers must decide how much to offer and how much to charge. To calculate the price that will maximize revenue, the monopolist firm needs to know the marginal revenue curve. To know the marginal revenue curve, it needs to know the slope of the demand curve. Without that information the monopolist has to make a guess about consumer demand at all the prices they can charge above the perfectly competitive price. The firm has to guess how the consumer will react to its price strategy.

Consumers have a similar problem. If they stop purchasing goods at the price the monopolist posts, they may force the firm to lower it. If they do purchase goods at that price, the monopolist may raise the price further. Both parties have a *strategic interaction problem*: their best strategy—what price to charge or whether to accept it—depends on the other party's strategy. In perfect competition everyone knows everything, including the fact that everyone knows everyone else's strategy. That's because there is only one strategy: take the price given by the market. When people can set prices, both parties have a strategic interaction problem. In this case it's one they can solve only by repeated experimentation—raising and lowering prices to see how much profit the monopolist can extract, and raising and lowering purchase offers to see how much power the consumer has to force the price down.

These strategic interaction problems never arise in perfect competition. But that means that if economics is going to provide a theoretical model of monopoly, it's going to need a theory of strategic interaction.

Every Market Is a Monopoly; No Market Is a Monopoly

Claim: There aren't any monopolies, or hardly any, in the real world.
Counterclaim: In the real world every firm is at least a little bit of a monopoly.

There's an argument to be made for each of these contradictory claims. Here's the support for the assertion that all sellers are monopolists to some extent: No two commodities, not even two quart bottles of milk in the supermarket dairy case, are absolutely identical. One is closer to your reach. If no two commodities are ever the same, then every market for every commodity is effectively a monopoly. If there are no perfect competitors for what you are selling, you have a monopoly to the extent of the difference between the competing products in meeting consumer wants. Therefore every market is a monopoly out there in the real world.

Here's the argument that no sellers are monopolists, at least not for long: For every commodity on the market there are substitutes—other goods and services that are reasonably satisfactory alternatives. If you don't like the burgers your local fast-food monopolist is selling, buy a pizza, or order Chinese takeout, or cook a meal at home. If no one is selling a substitute for some monopoly good, that's a business opportunity. Find a substitute and market it.

The real moral of this story is that like almost everything else in economic theory, monopoly is a model, an idealization of a phenomenon found in the real world to greater and lesser degrees. There is a spectrum of conditions that obtain in reality from something approaching perfect competition on one end to something approaching complete monopoly on the other end. The extreme states at either end of the spectrum are rarely if ever realized.

Economic Theory's War against Profit

Most markets seem to be closer to the perfect competition end of the spectrum most of the time. Gas stations may be local neighborhood monopolies, but they change their posted prices as their suppliers change the wholesale prices charged to them. They recognize that their monopoly power is slight.

Economists, especially new classical macroeconomists, have to believe that almost all real markets are far closer to perfect competition than monopoly. Think about the DSGE model that dominates macro theory. It's predicated on the claim that the economy is always moving toward its long-term, Pareto optimal general equilibrium. This is the state of affairs where all markets clear and all rents have been squeezed out of the economy. No rents, no monopolies, no deadweight losses, no consumer surpluses siphoned away as economically unwarranted profits. Monopolies can't be important parts of the economy that DSGE models describe.

Monopsony: A Harder Problem to Ignore?

Monopsony is the flip side of monopoly. It's when there is only one buyer on the market but many sellers. You might not think that monopsony is much of a real problem. After all, it's hard to name any markets that are monopsonies, except for major professional sports leagues—the NFL, the NBA, Major League Baseball, and the like. But as we'll see, monopsony poses a more serious problem for most of us than monopoly, even as it raises all the same problems for the economist's model of the economy.

There is one market on which almost all of us are sellers: the labor market. Only people who don't work for someone else can avoid selling on this market. So if this market departs much from optimality in ways that concede rents to the buyers of labor—employers—then most of us are likely to suffer from the inefficiency. And since the labor market is almost never competitive, we need to care about monopsonies and whether the labor market looks like one.

The model for monopsony (figure 7.3) is not as well known as the monopoly graph. It's usually drawn to analyze monopsony in labor markets like the NFL, where the only buyer of football player talent is a cartel of team owners working in collusion to run the NFL draft, assigning players to teams that they can take or leave.

Under perfect competition, the wage rate W_1 is set at the equilibrium where the labor supply curve (sometimes labeled ACL, the average cost of

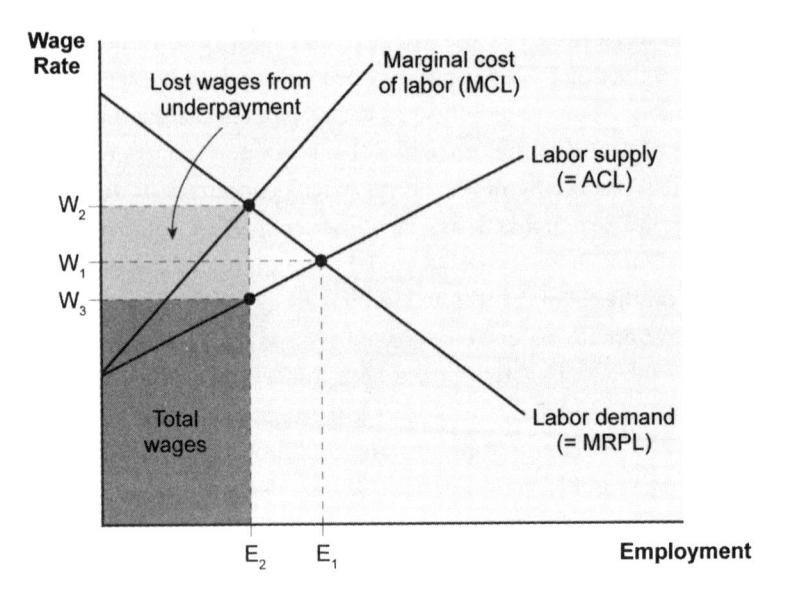

Figure 7.3
A monopsony graph.

labor) crosses the labor demand curve. The demand for labor is labeled MRPL, an initialism of the marginal revenue product of labor. This is how much it costs in additional labor to produce one more unit of the product to be sold. The lower the cost, the more labor will be hired. But if there is just one employer, there is no competition with other firms buying labor. The firm is not a price taker. It can set the wage where it wants.

The firm knows its marginal cost curve for labor (MCL) because it knows its costs per unit produced, including its labor costs. The marginal cost of labor curve rises more steeply than the labor supply curve because as the employer pays a higher wage to get more workers to work, it has to pay that higher wage to all the workers already at work. It also knows the values of the labor demand curve (MRPL).

Where the MRPL curve crosses the MCL curve tells the employer how much labor to buy—how many hours of workers' time to pay for. In figure 7.3 it's E_2 on the *x*-axis. Now follow the dotted line from E_2 to the labor supply curve. That's how the employer can find the wage rate that will get the amount of labor the firm needs.

Here's where market failure emerges—deadweight loss, rent seeking, and the exploitation of labor. The amount of labor that maximizes revenue for

Economic Theory's War against Profit

the employer, E_2, intersects with the supply curve for labor at W_3, which is below W_1, the wage rate at perfect competition. In fact, if the employer were to pay workers for the value they actually add, the marginal product of labor, it would pay the rate at W_2, which is even higher than the wage rate at perfect competition. But in fact the firm is paying less than the perfectly competitive wage.

In figure 7.3 the lighter gray rectangle above the dark gray rectangle reflects the amount of additional wages that the workers deserve for the quantity they produce, and the dark gray rectangle represents the amount they are paid. The narrow rectangle between W_3 and W_1 is the rent the firm makes by not paying the perfectly competitive wage. The deadweight loss is the triangle to the immediate left of the point where the perfectly competitive supply and demand for labor cross at E_1.

The same problems of measuring deadweight loss and the monopsonist's rent arise here. The amount taken from workers and the quantity of utility destroyed can be made to look large or small by changing the slopes and intercepts of the curves. But there is no way to actually determine the shapes of the curves and calculate the areas under them. All we can be confident about is that they exist if there is such a thing as utility, or something enough like it for the model to apply.

This is where another strategic interaction problem comes in. Just as in monopoly, there's a problem for both monopsonistic buyers of labor and the workers who sell it. The monopsonist doesn't know the shape, slope, or intercepts of the labor supply curve. The firm can't collect information on the factors that affect workers' willingness to work at various salaries. It doesn't even know what the wage rate would be under perfect competition, so it doesn't know how low a wage rate it can get away with. All it can do is guess, experiment with wage rates, and gauge whether they increase profits. Likewise, workers have to decide whether to work at the offered wage or hold out for a higher rate. These problems don't exist under perfect competition, where all parties are price takers.

Real Labor Markets Are Almost All Monopsonistic

Since most of us are sellers, not buyers, on the labor market, it's important to know where that market is on the spectrum from perfect competition to monopsony. The closer to monopsony, the more market failure, inefficiency, unearned profit to employers, exploitation of labor, and underemployment. From where most of us stand, it certainly looks like the market has lots of sellers of labor and relatively few buyers, giving employers real price-setting power. In the old company towns of the US Industrial Revolution, there was just one factory, mill, or mine. They've been replaced by Amazon fulfillment centers and Walmart superstores. But even where there is more than one employer, there are usually few enough that it is easy for firms to collude, either explicitly or by keeping track of one another's employment practices, wage rates, and fringe benefits. Smith recognized this tendency in *The Wealth of Nations*: "People of the same trade seldom meet together, even for merriment and diversion, but the conversation ends in a conspiracy against the public, or in some contrivance to raise prices."[1] Or in the case of the wages they pay, to lower prices!

It's easier to keep an informal coalition of employers together to act like a monopsonist than it is to keep a large number of potential employees together to act monopolistically, as labor unions hope to do. What this all means is that us working stiffs are more likely to suffer from employers' rent seeking in markets where we sell than from monopolies in the markets where we buy.

Modeling the labor market monopsony must begin with the stylized fact that in almost every such market there are many more sellers than buyers—more workers than employers. In fact, in many important markets the number of sellers is in the thousands and the number of buyers is in the single digits. Even in the markets for retail clerks and burger flippers, there are few buyers and many sellers. (The facts are stylized because we don't know the true numbers of employers and workers in most markets.)

It's pretty clear that the labor market in many places is close to monopsony. Walmart Supercenters in small rural communities provide apt examples.[2] In urban areas there are more buyers of labor than in rural areas, but the number is still usually an order of magnitude smaller than the number of sellers. Moreover, many sellers have to restrict their offers of labor to a small fraction of the total number of buyers. Consider waiters, sales clerks, fast-food workers, truck and taxi drivers, and others who can't work remotely. The effective market for their labor will be geographically limited in the same way the effective market for buyers of commodities is limited to a small number of sellers. And in contemporary economies, the destruction of retail markets by internet-based sellers such as Amazon or Uber further

Economic Theory's War against Profit

reduces the ratio of buyers of labor to sellers. These firms' sales destroy locally competitive retail sellers that may buy labor more competitively and further reduce the ratio of sellers to buyers in the labor market. In engineering, law, medicine, and other professions, the number of buyers will always be smaller than the number of sellers. This ratio will only get worse with remote working and online vacancy postings. Job postings by a single buyer of labor will reach almost everyone in what Marx called the reserve army of labor.

Close to the extreme end of perfect monopsony—just one buyer—is a much more common state of affairs: a small number of very similar firms, all seeking the same kind of worker at the lowest wage rates they can pay. These firms face a *collective action problem*. Once they have estimated the most advantageous wage rate they can jointly set, each firm has an incentive to raise wages just enough to scoop up all the best workers. How do they solve the problem of keeping the coalition together? It's a stylized fact well known in game theory that a small number of players regularly interacting with each other can remain in a long-term coalition without defections. We'll see how in the next chapter when we discuss public goods and the collective action problems they create for perfectly competitive markets.

The labor market is the market furthest from perfect competition and most like a monopsony. It's one in which the buyers have market power—the ability to be price setters—and in which almost all buyers are securing rent, or returns above what they would receive in perfect competition, which are reinvested in other market-failure-producing activities.

When Labor Market Reality Collides with Economic Theory

The perfectly competitive market chugs along nicely without money, bans profit, and so makes entrepreneurship impossible. It also has to make unemployment disappear, since all markets always clear, including the labor market. If there are people looking for work, wage rates will keep falling, making it more attractive to hire workers, until there are no more workers to hire. Recall Robert Lucas and Thomas Sargent discussing the discipline imposed by optimizing on new classical economists' conception of the economy after the wreckage of Keynesian economics had been cleared away.

But how to reconcile new classical economic theory with the occasional reality of persistent, decades-long unemployment? Could monopsony be part of the explanation? Not over the long term. Like monopoly, monopsony should be *self-liquidating*: employers' short-term rents should serve as signals that attract other potential employers into their sector to steal their workers with higher wages while still making (slightly smaller) rents until the short-run monopsony dissolves into long-run perfect competition.

The new classical economists had a better idea for explaining unemployment. They explained it away as a friction that can't be entirely eliminated, like Milton Friedman's natural rate of unemployment. In fact, the friction they described might be the source of Friedman's. Inspired by Friedman's treatment of the Phillips curve over the short and long terms, his permanent income hypothesis, and the rationality of expectations that make the money illusion impossible, the new classical macro models made long-term unemployment impossible all over again.

Could unemployment be an effect of short-term departures from long-term optimality, always on its way to disappearing at the general equilibrium? Nobel Prizes were won by economists for showing just that: by imposing RCT on buyers and sellers in a labor market they explained the existence of a little unemployment—the natural rate—as the optimal outcome of a labor market operating as efficiently as all other markets.

The way new classical macroeconomics treats unemployment reminds one of the adage that when your only tool is a hammer, you'll want to treat everything as a nail. To begin with, some of the unemployment that shouldn't exist in the economy will be short-term results of what the New Keynesians called sticky prices and explained (away) as long-run efficiency wages. Instead of cutting wages in a downturn, employers will keep their best workers at higher wages and "let go" (a great euphemism) less productive workers, whence unemployment. Remember that New Keynesians aren't Keynesians. They endorse the existence of general equilibrium in the long run. New Keynesians provide excuses for the late arrival of equilibrium in the temporary stickiness of prices and other briefly mistaken expectations.

Sticky prices aren't the whole story new classical economics wants to tell. Here's the rest of it—a story that won the Nobel Prize for its tellers, Peter Diamond, Dale Mortensen, and Christopher Pissarides. They constructed a model that turns unemployment into another friction that temporarily slows the economy on its way to optimal general equilibrium.

It works like this: Even though the labor market is as perfect as all the others over the long term, there is a short-term *matching problem* that workers

Economic Theory's War against Profit

and employers have to solve. It takes time to solve, and that's where unemployment comes from.

Employers set wages at the *marginal productivity of labor*. Everyone doing the same job gets paid the amount it costs in labor to produce and sell the last unit whose price equals its production costs. That's because workers and employers are price takers. If you're an employer and you lower the wage below the market price, everyone will quit. If you raise wages above the market price, you'll soon go out of business. If you're a worker and try to hold out for a higher wage, you won't be hired. Working for less than the market wage is crazy. But finding the right job, where the wage rewards your most productive skill, may take time and guesswork. On the employer's side, finding the worker who has the right package of skills may also take time. This is the matching problem in the labor market—how the right workers and the right employers can find each other. While they're both looking, there is a certain amount of unemployment and a certain number of job vacancies—frictional unemployment.

Enter RCT rationality. The person with labor to sell receives an offer. The seller has to decide whether to accept the price offered or to forgo it and continue to search for a higher wage. Workers—sellers of labor—have to guess what other jobs at other wage rates exist in order to decide whether to take the current offer. Each unemployed person has an estimate of the distribution of job openings at various salaries on the *x*-axis in figure 7.4. Workers are price takers who cannot shift the job and wage distribution curves. Employers, too, are price takers who can't change the curves.

If jobs are randomly distributed along a nice bell-shaped curve whose mean is at E(W), there are few openings at very high salaries and few at very low salaries. The largest number of vacancies is at some in-between value. Of course, there is some minimum wage level beneath which the seller of labor will not accept any job. What that is will depend on the worker's reserves or other sources of income. Call this minimally acceptable wage W_R. Above this is a wage at which the worker will immediately accept the job instead of continuing to look for one at a higher wage. Call this W_U. The closer W_U is to W_R, the minimally acceptable wage rate, the sooner the worker will accept an offer. The further above W_R that W_U is, the longer the worker will be prepared to decline the offer and search for a higher-salary job. Figure 7.4 illustrates these wage levels and their distribution.

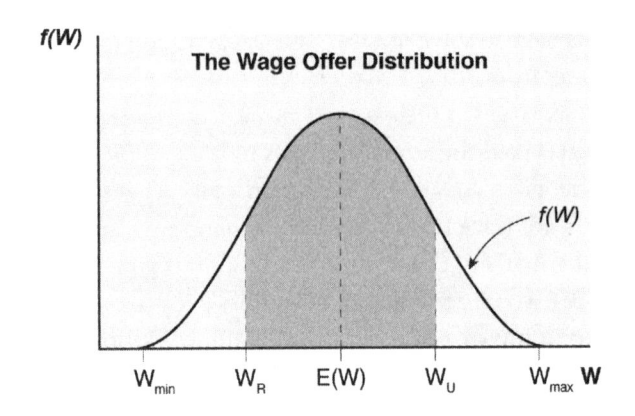

Figure 7.4
The worker's RCT choice problem.

Sellers of labor engage in a search during which they calculate the cost of waiting against the benefit of a higher-wage job whenever they have a job offer. Employers have a similar task. They are always either losing some workers they want to retain but can't pay enough to keep or firing workers whose wages exceed their marginal productivity. Employers know the minimum skill level needed for available jobs, the distribution of skills and abilities among sellers of labor, and the wage rate they must pay everyone doing the same job. Employers have to sift through job applicants till they find a worker above the minimum satisfactory level willing to work for a wage at or below some maximum above which the firm loses money. The delays in employment brought about by the combined searches of buyers and sellers of labor produce a steady level of unemployment over the long term, even in an economy at general equilibrium.

By combining the models of employer search and worker search, economists can derive some stylized facts: The higher the social safety net provided to unemployed workers, the longer the duration of their job search and the fewer jobs filled per unit of time. Productivity changes that improve the marginal productivity of labor reduce the duration of unemployment and also reduce vacancies by shifting the wage curve to the right.

But there is always some number of vacancies and some level of unemployment, and there would be even under perfect competition, because of the friction introduced by the search. This, according to new classical macroeconomists, is the source of unemployment.

Let's go back to the prehistory of new classical economics. Recall the argument against the Keynesian policy of reducing unemployment by increasing inflation that Friedman mounted even as it was still guiding employment policy in the 1960s. Friedman's argument introduced the concept of a natural rate of unemployment the economy couldn't go below. It was the vertical line of the long-term Phillips curve (figure 6.1) and portended the use of the term *natural* in the DSGE models of chapter 6 to refer to the economic general equilibrium at which all markets clear and there is a Pareto optimal distribution. Friedman's natural rate of unemployment reconciled that blessed state with the existence of some unemployment.

Unemployment can't be brought down to zero, ever. It was the original friction in the new classical economists' macro models (to which the financial crisis of the 2000s added several others). With the stylized fact of a natural rate of unemployment, economists could go on to enrich their explanation of the Phillips curve and governments' inability to affect unemployment by changing the rate of money-price inflation. If the unemployment level were pushed below the natural rate, the results would be inflation and a temporary reduction in unemployment, but since money is neutral and people have rational expectations, unemployment would soon go back up even as money prices increase, whence stagflation—the "wreckage" left by Keynesian macroeconomic policy.

Economists uncomfortable with the word *natural*, owing to its connotations of moral approval, substituted a more technical term for Friedman's natural rate of unemployment: the non-acceleration inflation rate of unemployment, or NAIRU. Friedman never offered a number to describe his natural rate of unemployment. It was just a stylized fact used to explain another stylized fact—the short- and long-run Phillips curve.

The search model of unemployment as a frictional effect of the matching problem facing rational workers and employers gave us the NAIRU that new classical macroeconomics needed. In the search model of unemployment everyone is still a price taker. Neither employers nor workers can make salaries budge because they are set by the market-clearing price of labor at equilibrium. There is no strategic action problem to solve.

It's also worth noting that the search model of unemployment treats what's going on in people's heads as the causes of economic outcomes. It can't be thought of as just an idealized model of unemployment trends that has nothing to do with human psychology—how individual people calculate their choices. The search model completely belies the way Friedman and Gary Becker, back in the late 1950s, defended economic theory's lack of interest in whether individual buyers and sellers were really rational decision makers or not.

Economists are stuck with the search model of the labor market because giving it up would unravel so much of what they have constructed since the demise of Keynesian macroeconomics. The trouble is that wage rates don't obey the parameters of the market-clearing frictional-unemployment-only state of affairs the model requires. Economists have noticed that the search model has no way of accounting for large observed variations in unemployment, the cyclical character of changes in job vacancies, or wage rigidity. Then there is the fact that large employers pay better, on average, than small ones. New classical modelers have spent the two decades since the frictional match model was formulated trying to shoehorn these and other "anomalies" into it: frictions added to frictions.

In a way the last few pages have been a digression from our treatment of market failure, but an insightful one. Unemployment is probably the most important public policy issue economics has dealt with since the advent of active government management of economies. The availability of explanations for unemployment that turn on market failure—for example, monopsony—hasn't deflected economic theory from the models of perfect competition that make substantial, persistent, long-term unemployment not just inexplicable but impossible.

In the next chapter we'll come back to monopsony in the labor market and why that particular market failure won't be unraveled by the mechanism of the competitive market once we've acquired the tool kit of game theory.

Externalities: Usually Negative, Rarely Positive

There is more than one way sellers can impose their prices on others. In doing so, they again shift the economy away from perfect efficiency in its output and secure unearned rent. And because they are often monopolists or stable coalitions setting prices together, their market power increases still further.

The clearest and most important case of this sort of market failure is pollution. When a local factory pumps smoke into the air, dumps waste in the river, buries toxic chemicals near residential districts, runs its loudest machines all night long, and sites buildings where they block neighbors' views of distant snowcapped mountains, economists call those activities *negative externalities*. When the factory owner plants a flower garden in front of the main entrance so they can enjoy it on the way to visit their factory, all passersby can enjoy it too, without charge. That's a *positive externality*.

Environmental economics is a growth area of the discipline, one preoccupied by the costs imposed on all of us by damage to the environment that is not paid for by those who bring it about. The market failure resulting from the imposition of pollution externalities can be huge. Climate change is a negative externality we all suffer from, and the costs it imposes deprive us of many other economic opportunities. Environmental economics is another domain in which the theory is limited to identifying and explaining stylized facts, but ones of the greatest importance. An off-the-shelf analysis of the externalities of pollution shows why mainstream economic theory can't really deal with the problem.

A negative externality is a cost of production that the producer of a commodity doesn't pay for but forces others to pay. This enables the producer to sell at a price below production cost, which enables the firm to make rents. The cost imposed on others is often environmental damage the firm gets away with not paying. I'll use, as an example, a firm producing and selling steel.

Economic theorists' approach to externalities starts with another graph rather like the ones that take apart monopoly and monopsony to show how these forms of rent-seeking deform the optimally efficient economy (figure 7.5). In figure 7.5, the dark gray triangle A represents the deadweight loss. The largest trapezoid to its left, on the other side of the demand curve D (which includes the smaller shaded trapezoid B and the larger one, C), is what bystanders unwillingly pay to subsidize the firm's production. It's similar to the consumer surplus in a monopoly. Where does this large trapezoid area (B+C), the consumers' loss, come from? D is the demand curve for steel per ton, and S is the firm's supply curve. Assume its supply curve tracks its marginal cost of production. The demand and supply (or marginal cost) curves cross on the right side of the graph where price is P_s and quantity is Q_s. But suppose that for every ton of steel produced, the firm does $10 worth of damage to the local air quality. That must be added to the marginal cost, since someone is going to have to pay it. So the real

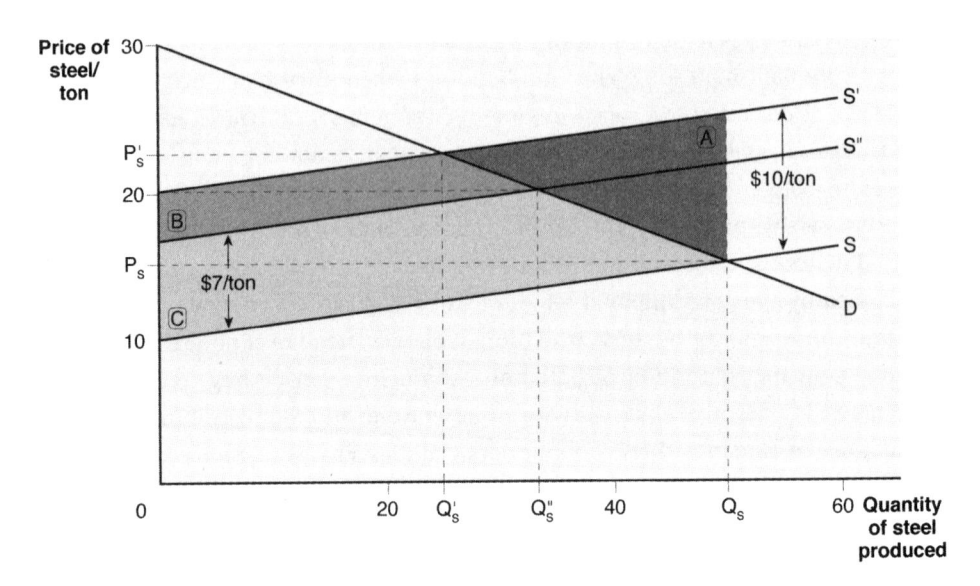

Figure 7.5
The negative externality problem.

marginal cost curve is S', which means the firm is selling below the price required by an allocatively efficient market. As the graph shows, the price should be $10 more per ton at every quantity produced. At this price, the demand curve meets the higher marginal cost curve at Q_s', well to the left of Q_s. At that price, P_s', and that quantity, Q_s' (which includes $10 per ton paid to neighbors for the right to pollute their air), the firm's production is allocatively efficient. Suppose the steel company spends only $7 per ton to clean up the air. That still leaves the shaded area B as the neighbors' added cost of dealing with dirty air. B is the same shape as the original lost consumer surplus, but smaller because the polluters are ponying up a part of the cost of the externality. It is the amount of externality still imposed on the neighbors by the steel company that needs to be internalized if the economy is to attain an allocatively efficient outcome.

It's a neat graph, but like the monopoly and monopsony graphs, it hides some serious problems that face the environmental economist using off-the-shelf theory to analyze environmental damage and decide who should pay for it. First we need to ask how to calculate the actual cost of air pollution damage to the neighbors. The truth is that in most cases it's a made-up number. The only way to price the damage "correctly" would be

Economic Theory's War against Profit

to have a perfectly competitive market in air pollution damages. Then the neighbors could decide how much pollution damage to accept in exchange for some amount of money. But there are few markets in which people can exchange environmental harms for money, let alone perfectly competitive ones.

Measuring the consumer surplus and deadweight loss is even harder. It's not just that these are stylized facts about welfare losses in incalculable units. Assume that the polluting steel company can undercut other steel producers using the same production processes by $10 per ton. Can we infer that the externality imposed on neighbors is equivalent? We know that the polluting steel company's rent could be used to cut costs and drive other companies out of business, after which the monopoly firm would have a problem figuring out how much to charge for a ton of steel. But meanwhile, where did the additional $10 per ton figure that responsible companies are paying come from? It wasn't set by a competitive market transaction. We're back where we started.

The great problem facing the environmental economist is that economic theory is not much help in identifying the real costs imposed by the externalities it is supposed to price. The externality analysis that economic theory offers us establishes the existence of a stylized fact. But it doesn't have the resources to take us beyond stylized facts. We need quantitative measures so that environmental policy analysts can frame regulations and laws controlling pollution and lawyers can file scientifically substantiated damage claims against polluters.

Trying to Make Market Failure Go Away by Squinting

Sometimes you can see things that are far away by squinting. Economic theorists tend to try to see past market failure by squinting, too, looking further into the long run to make the failures appear smaller and less important.

One reason the theory of perfect competition focuses on long distance is that it doesn't have the tools to look closely at what is staring it in the face. Economic profit, rent, entrepreneurial successes—the "creative destruction" of nicely behaved markets—are the stock in trade of real-world economies. We've been seeing them in action up close at least since the Industrial Revolution. But for economic theory they have always been blips, aberrations, exceptions, and blemishes, departures from the natural state of things that are overwhelmed by the normality of the long-term general equilibrium, where well-behaved probabilities rule and money has no causal role.

When we stop squinting into the distance and just glance at the short term, when we are all still alive, the idealizations required for perfect competition no longer look harmless. In fact their departure from reality—or rather, reality's departure from them—is so great that there is price setting, rent seeking, economic profit, and market failure *everywhere*. And that means the theory descended from Adam Smith is not just a blunt instrument. It doesn't work at all. We need a new theory, one that can cope with price setting. What economics needs is a theory of strategic interaction. What it needs is *game theory*.

Game theory is a terrible name for the most important theory in the social and behavioral sciences. Game theory is not about games, at least not literally. Some games illustrate in starkest terms the thing that game theory is really about, which is how it got the name. What almost all games have in common is that how the players choose to act depends on their calculation of the other players' strategies. Most games confront their players with such strategic interaction problems: the need to figure out what the other person is going to do and what the other person thinks you are going to do in order to choose your best strategy.

Strategic interaction problems arise in every case of market failure. It's a phenomenon that we can understand only by using the tools of game theory. And since the real economy is nothing but market failures—sometimes tiny, sometimes small, sometimes large, sometimes immense—game theory is the most important theory there is for the economist.

This chapter began with the observation that there is no more room for profit than there is for money in the economic theory driven by Smith's conjecture and the proof of the first theorem of welfare economics. The fact that both money and profit are very much with us in economic life is a primary indication of the need for some other approach to the economy than the theory that makes both money and profit into short-term aberrations.

8 It's All Game Theory Now

No market is perfectly competitive; no market is even close. Market failure is everywhere, and it's almost never decreasing and almost always increasing. It has to. We'll see why in this chapter.

The price setters who drive market failures to secure their rents all have the same strategic interaction problem: deciding how much to charge or to pay. Their problem imposes strategic interaction problems on us, their price takers, who have to take their prices or leave them. Therefore, what economists need in order to understand economic behavior is something quite different from the tools provided by the model of perfect competition in which everyone is a price taker. The economist needs a theory of strategic interaction.

The labor market and financial market are the two most important markets for almost all of us. The labor market is the only one in which almost all of us are sellers. The financial market is the one we rely on to make our future lives better through economic growth. They are the two biggest domains of market failure, price setting, market inefficiency, and persistent and almost unstoppable increases in such problems.

It's the structure of these two markets that makes departures from optimality persistent and ever increasing. What does that mean for economics and the economy? The problems created by the structure of these two markets are the reasons capitalism is the worst of all economic systems *except for all the others*. The job of economic theory is to provide us the tools to understand how to minimize the damage.

The Prisoner's Dilemma and the Supply of Public Goods

One way to introduce game theory and see how it helps is via its most famous model: the prisoner's dilemma (PD), which is known even to people who have never thought much about economic theory.

Imagine this: You and I are partners in crime, but we are nabbed while carrying our safecracking tools just outside the jewelry store we plan to rob. The cops separate us and tell each of us exactly the same thing: If both of us confess to attempted robbery, we'll get five years each. If one of us confesses and the other doesn't, the fink will go free and the guy who clammed up will get ten years. If neither of us confesses, we'll both get two years for carrying burglary tools.

Here's a diagram, called the payoff matrix, that summarizes your situation and mine. There are four boxes, marked I through IV. My jail time is given in the lower left of each box and yours in the upper right.

	You Stay silent/don't defect	You Confess/defect
Me Stay silent/don't defect	2 years in jail / II / 2 years in jail	0 years in jail / III / 10 years in jail
Me Confess/defect	10 years in jail / I / 0 years in jail	5 years in jail / IV / 5 years in jail

Assume all you care about is minimizing your time in the slammer. The same goes for me. What should each of us do? I reason this way: If you confess, I'd be a fool to stay mute, since I'd get ten years. If you don't confess, then I've got a chance to go free if I confess. So either way, I'm going to confess. You reason exactly the same way. We both confess, ending up in box IV. But wait—we could have done much better if we had both stood mute. Then we'd only get two years each instead of five. The trouble is that the more confidence I have that you'll keep your trap shut, the stronger my incentive to confess, and the same goes for you. If all we care about is minimizing jail time, there is no way we can get into box II. That's what makes this situation a dilemma: rational agents are stuck in box IV even when they'd both prefer to be in box II. You and I can't get there (box II) from here (box IV).

This is the structure of a prisoner's dilemma in game theory:

My preference ranking: I > II > IV > III
Your preference ranking: III > II > IV > I

Every game in which both players prefer box II to box IV but can't get there is a prisoner's dilemma.

The payoffs don't have to be less or more jail time; they can be any amounts of anything people want more or less of, so long as the preference rankings are as above. Here's another example, in which two people are faced with quite different payoffs but it's still a prisoner's dilemma. We each have two cards, marked C for cooperate and D for defect. We put them out simultaneously. If we both play C cards (cooperate) we get the amounts in box II. If we both play D cards (defect) we get the box IV payoffs, and if we play different cards from one another, we get the box I or box III payoffs.

		You	
		C card	D card
Me	C card	$64 II $50	$65 III $25
	D card	$63 I $100	$62 IV $47

My preference ranking: I > II > IV > III
Your preference ranking: III > II > IV > I

It's still a prisoner's dilemma, even if the payoff amounts are different, because the preference rankings of two rational players are the same as in the jewelry heist scenario. Games are defined by the preference rankings of the players.

Coupon wars between local businesses are a real-world example of PD strategic interactions. The local Papa Johns and Domino's franchises can each issue coupons or refrain from doing so. If one gives discounts and the other doesn't, the coupon-issuing store will do much better. If both give coupons, they'll both do much worse. What to do? Apply the reasoning from the jewelry store heist PD. If Papa Johns doesn't give out coupons, then the best strategy for Domino's is to give them out. If Papa Johns does give them out, Domino's has to, too. Either way, they are stuck in a stable equilibrium in which both are worse off than they could be.

Pizza-Coupon War

		Papa Johns	
		No coupon	Coupon
Domino's	No coupon	$5K II $5K	$10K III $1K
	Coupon	$1K I $10K	$2K IV $2K

The dollar amounts are increased revenue per month. The best solution is collusion between the franchises to not issue coupons, which Adam Smith said they'd always find a way of doing: "People of the same trade seldom meet together, even for merriment and diversion, but the conversation ends in a conspiracy against the public, or in some contrivance to raise prices."[1] But once they've solved this strategic interaction problem, two others immediately arise. The first new problem for each of the colluders is to monitor and enforce the (often unspoken) agreement to collude. The second is to seek some way of breaking out of the agreement that the other party can't respond to quickly enough. Without a way of sidestepping the two new problems, the collusive solution to the first one unravels.

Nobody wants to be in a PD because there is no way to get from the less preferred outcome into a more preferred outcome. But sometimes it can't be helped. The problems for economic theory raised by *public goods* illustrate how significant the prisoner's dilemma can be. It creates one of the most serious kinds of market failures.

Public good is a technical term in economics. It doesn't mean something provided universally and freely by the government. (Almost no goods delivered to the public by the government are what economists call public goods.) As the economist defines the term, a public good has two very special features: First, one consumer can get some only if other consumers can, too. Even if one person pays for it all, they can't stop other people getting some of it without paying. A public good is *non-excludable*. Second, there is no way for any one buyer to consume the good without leaving as much to others to consume as they want. Consumers aren't rivals in the consumption of a public good.

You will immediately wonder how there could be such a good. The nicest example of one is street lighting in dangerous neighborhoods. You can't consume the increased security that street lights supply unless others do, too. If you pay the electricity bill for the streetlights, the only way to stop others who haven't paid from enjoying the benefit is to turn them off—but then you won't be able to consume what you paid for. Their consumption is therefore non-excludable by your consumption. Additionally, no matter how much security from street lighting you consume, anyone and everyone else can still consume the same amount. Consumption is non-rivalrous. Street lighting is a public good. Of course, different people will get more or less out of street lighting. Women probably secure greater benefit than men, and people who don't go out at night don't get any.

The problem public goods raise is that no matter who pays for them, everyone can consume them. But that means almost no rational agent would be willing to pay any amount, no matter how small, for public goods. They'll get them so long as someone else wants them enough to pay the whole cost. Everyone reasons this way, so no one puts up the money, and everyone is worse off than they could be or want to be.

An example shows why the provision of public goods generates a prisoner's dilemma. Consider Alf, who, like a thousand other people in town, must keep air pollution abatement equipment on his car, even though it reduces engine performance and increases his maintenance costs. Of course, he greatly prefers the benefits of clean air to the benefits of better performance for his car at lower cost. But clean air is for all intents and purposes a public good: Alf gets as much as he wants everywhere he goes without depriving others of it, and he couldn't exclude anyone else from consuming the clean air if he tried.

But Alf is a rational agent who reasons that if he disables his air pollution abatement equipment while everyone else keeps theirs on, he'll have the best of both worlds—clean air and cheaper, more fun driving. He also recognizes that if everyone or even most people detach their abatement equipment, the air will be bad no matter what he does. He'd be a sucker to keep the equipment on his car if a lot of other people weren't doing the same. The rational thing for him is to detach the device, no matter what other people do. So Alf finds himself in box I, while everyone else in box I with him is still better off because everyone enjoys the clean air.

Should Alf Disconnect His Air Pollution Abatement Equipment?

		Enough others	
		Keep device on	Disconnect
Alf	Keep device on	Good air – performance II Good air + performance	Poor air – performance III Poor air – performance
	Disconnect	Good air – performance I Good air + performance	Poor air + performance IV Poor air + performance

This doesn't look like a PD based on the preference rankings:

Alf: I > II > IV > III

Enough others: I ~ II > IV ~ III

The tilde indicates indifference, or no preference between the alternatives. Alf and everyone else prefer box I, where everyone gets the public good (clean air) while Alf also gets better, cheaper performance for his car.

But then Betty looks at the same matrix and sees it's rational for her to disconnect, as Alf did. Then it's Charlie's turn, Dawn's, Ed's, and Flora's, until everyone disconnects and everyone is in box IV without a way to get to box I or box II anymore. The provision of the public good unravels entirely into a multiperson prisoner's dilemma.

The general case looks like this:

		Enough others	
		Contribute	Withhold
Alf	Contribute	Good II Good	Poor III Worst
	Don't	Good I Best	Poor IV Poor

Alf: I > II > IV > III

Enough others: I ~ II > IV ~ III

Thus the provision of public goods is another case of market failure. The perfectly competitive market won't just fail to provide an *optimum* supply of a public good. If people are truly rational, a perfectly competitive market won't provide *any* amount of any public good, with one exception: when one consumer wants the public good so much they are willing to pay for it all. Then it becomes a positive externality provided to everyone else as a side effect. Think of the garden a rich family plants in their front yard. Passersby can enjoy the sight non-rivalrously and non-excludably, until the family puts up a fence to prevent littering on their property and theft of their flowers.

How many public goods are there? How important are they? Are there public goods besides street lighting—things none of us can enjoy without allowing others to share, that none of us can gobble up till there isn't enough left for others? Street lighting shares its public good status with the presence of police to deter vandalism, theft, and other crimes. Then there is fire protection: in densely built urban areas I can't buy it without also protecting my neighbors' houses from destruction by fire, so they can free ride. National defense was offered as a powerfully attractive public good during the Cold War. Maybe it was; public goods don't have to be of the same value to everyone. A crucial public good is now undersupplied due to free riding in a prisoner's dilemma: keeping the average world temperature from rising more than two degrees Celsius. It's undeniably a public good, and the failure of perfectly competitive and real markets to provide it is a critical market failure.

Many private goods produce public goods as side effects. Public schools are not public goods: Your kid can be excluded by being suspended or ruled ineligible because you don't pay real-estate taxes in the town. Your kid's consumption of teachers' attention is rivalrous; the more of their time your kid takes, the less there is for other kids. But having a good school system provides spillover public goods to everyone in town. It attracts the sort of people whose business or quality of life benefits from good schools even if they have no children of their own.

Traffic laws are another example of public goods that are consumed by motorists. In fact, any rule of law is a public good. Even the worst laws are preferable to complete anarchy, Thomas Hobbes's "state of nature." They provide a non-excludably and non-rivalrously consumed minimal degree of predictability in human interactions. This is one way game theorists have reconstructed Hobbes's argument in *The Leviathan*. Hobbes implausibly argued that giving all our power to a sovereign was the only way out of the state of nature, the "war of all against all."[2] If the rule of law, any law, including the arbitrary rules of an absolute tyrant, are public goods (i.e., better than the state of nature), then there is something to Hobbes's argument.

Notice that the problem of providing public goods is like the pizza franchises' prisoner's dilemma, only harder. Once they've agreed to refrain from offering coupons, the two franchises have to find a way to make their collusion last, even as they look for ways to cheat on their agreement. When it's just two stores, the collusion-monitoring problem isn't so hard to solve. Both have credible threats to keep it in force. But when providing the public good of clean air requires not only Alf but Betty, Charlie, Dawn, and a hundred or a hundred thousand others first to agree and then to monitor their agreement, the repeated or iterated strategic action problem becomes much greater. Game theorists have from the start emphasized the *size problem*—the bigger the coalition needed to provide the public good, the larger the transaction costs of securing agreement among everyone involved and the harder the problem of enforcing agreement.

We'll come up against these two related features of the market's failure to produce public goods over and over again in the following chapters. They constitute collective action problems. These are the most important problems faced by working stiffs like us. We have to solve them if we are to escape rent seeking that endlessly increases market failure to firms' advantage. Meanwhile, our treatment of public goods gives economic theory the resources to identify an urgent problem facing both real economies and the optimality argument for the model of perfect competition.

Good Ideas: Another Market Failure

Maybe you don't think public goods are very important. But there is a commodity that is almost a public good and is of the utmost importance to every economy, culture, and civilization: good ideas. By *good idea* I mean an idea that pays: a way to do things better, more cheaply, or differently that makes people better off by increasing efficiency or productivity or supplying people with something they didn't even realize they would want. Consider the electric light, or Henry Ford's innovation of making things on an assembly line. An idea can be almost a public good, and therefore never likely to be supplied at an optimal level by a competitive market.

Here's a good idea that has been around for a long time: rotating crops to increase annual agricultural yields. But rational agents would never have invented or perfected it. To begin with, any rational agent who tried it for the first time would be taking a huge risk. What if crop rotation didn't work? They'd be chancing their family's food supply over several years. The only way we can expect this good idea to emerge is by serendipity. More importantly, no matter how one farmer hits upon the idea, once other farmers see how well it works, they can adopt the same good idea without depriving the original inventor of any amount of continued use of it. Good ideas are consumed non-rivalrously.

The discoverer of a good idea has an incentive to keep the idea secret, in this case to exclude others from implementing crop rotation. With higher yields, suddenly the farmer is making rents. But how to keep the idea secret? The farmer could build a high wall around the fields, or start rotating crops only in fields far from prying eyes. But this is costly and fantastically inefficient, and it probably won't work for long, anyway. People have an incentive to find out why their competitors are doing better than they are. Crop rotation is very difficult to make excludable. Most profitable ideas are non-rivalrous and difficult to exclude others from using. They are enough like public goods that the free-rider problem applies. Good ideas won't be supplied by a perfectly competitive market at anywhere near the optimal level, if at all. That's bad news for the prospect of growth or productivity improvement in the economy, and bad news for the perfectly competitive model—it has no room for money, no scope for profit, and now no possibility of productivity-driven economic growth. We can bid farewell to the hope that the perfectly competitive model can be reconciled with short-term real profits driven by entrepreneurial innovation. If we are going to make room for innovation in the world that economic theory describes, it will have to be with a different economic theory.

This is where game theory steps in to help design a solution. What we need is a new institution, some substitute for the competitive market, that will give people an incentive to invest in imagining the good ideas we all need. No amount of pushing the real market toward perfection will solve this problem. We need what economists call a *non-market second-best* arrangement to get good ideas flowing. It's called *second best* because in economic theory, the perfectly competitive market is best due to its optimality. Second bests give up optimality as an unreachable ideal.

The standard second-best solution to the problem of incentivizing good ideas is the patent. Designing the government's patent rules was a game-theoretical exercise in institution design long before the name *game theory* was coined.

The patent is a monopoly given in an exchange. The government trades with the inventor, conferring a legal monopoly on selling the use of their idea for a certain number of years in exchange for a complete public description of how the idea works and what it's good for. Notice that granting the monopoly is already giving way to a market failure. The trade minimizes its harmful effects. This package—monopoly and full disclosure—is designed to maximize incentives to invest in the discovery of good ideas and to maximize efficiency in the implementation of good ideas throughout the economy.

There are standard caveats on patents. First, the idea has to be immediately useful. You can't patent basic scientific discoveries (which usually don't have immediate paying applications, in any case). Second, you can't patent good ideas that are already in circulation—no patents are to be awarded for reinventing the wheel. The information disclosed by the inventor allows everyone to figure out whether they can profitably make use of the idea. That way everyone can figure out whether it's worthwhile to pay the patent holder's monopoly price. Monopolist patent holders have to figure out prices that will maximize their rents—if a price is too high there will be no buyers, but if it is too low the patent holders will lose out on potential income. Setting the price is the patent holder's strategic interaction problem.

How long should monopolies on good ideas last if we want to maximize the rate of invention and discovery? There is no fixed answer to that question. Governments have to experiment with different lengths of monopoly time to find the optimal one. Whatever the time allowed, history has shown that inventors will try to extend their monopolies by devising "improvements" on, "new uses" for, or new ways of making the product. There is an entire body of patent law devoted to finding ways around patent laws.

Inevitably, there will be an arms race between patent holders and the government, acting for the whole economy and seeking to minimize the deadweight loss and consumer surplus reduction that the monopoly imposes. The government will tweak the patent rules and the patent owners will respond, seeking to continue securing their rents.

Game theorists therefore search for alternatives to the patent system that could produce a better second-best solution. The Nobel Prize was awarded in 2018 to an economist, Michael Kremer, who proposed a way to preserve more of the consumer surplus destroyed by monopoly patents. Kremer's idea makes use of auctions, which game theorists have studied a lot, as we'll see in chapter 10. Here's an example of his innovation for pharmaceutical patents—probably the most important good ideas to be made affordable. Say a pharmaceutical company patents a new drug. The government offers to buy the patent at, say, 10 percent above the value set by an auction in which other firms bid for the patent. The bidding reveals expert estimates of the market value of the drug. The patent holder can accept or decline the high bid plus the government's bonus of 10 percent. If the firm accepts, then the government owns the patent and in a random 10 percent of cases, the government sells the good idea to the high bidder. This gives every company an incentive to bid their best estimate of the drug's value. In the other 90 percent of cases, the government buys the good idea at the high bid plus 10 percent and gives it to everyone for free. Then the pharma industry competes to produce and sell the drug. If the 10 percent chance of being able to buy the patent isn't enough to induce other firms to bid in the auction, it may have to be raised. If there are many bidders, then the 10 percent chance can be lowered. Where to set it is a strategic interaction problem.

Kremer has argued that this more complicated patent system is a better second best that brings the drug development market closer to optimality. Whether it ever gets implemented or not, Kremer's idea is a fine example of institution design driven by game theory. We'll encounter more examples in chapter 10.

Solving the Public Goods Problem without Coercion

Once game theory and the analysis of market failure found each other, the obstacles to making real markets closer to perfectly competitive became much more apparent. They are almost always problems of collective action: convincing economic agents to coordinate their choices when the temptation to free ride is present and the costs of cooperation are high.

Mancur Olson, an economist, identified the size problem early on. The larger the group, the higher the transaction costs of agreement, payment, enforcement, detection, and punishment of free riding. The more people that have a stake, the larger the space needed to get them together, the more complicated the system for collecting payment and forcing people to pay, the more difficult the identification of cheaters, and the more expensive the adjudication of cheating and the imposition of punishments for doing so. Large enough groups of rational agents could never solve their own collective action problems—certainly not with unanimity or without coercing free riders.

This is the game theorist's explanation for the persistence, if not the historical origin, of government among rational agents. Ever since Hobbes, it has been recognized that one way—perhaps the only way—to assure the provision of public goods like the rule of law is to give the government the responsibility for producing the public good and the sole right to employ violence to extract payment from all for the public goods that all consume. Government is the solution to our collective action problem. But it's obviously an imperfect one. Nondemocratic governors almost always abuse their monopoly on the legitimate use of force to benefit themselves. In a representative democracy, elected officials oversupply public goods and under-coerce voters to secure the resources needed to pay for them. That way they win reelection.

It took a political scientist, Elinor Ostrom, to show how groups in the real world provide themselves with public and collective goods without governmental coercion. Ostrom won the Nobel Prize in economics for doing so (though some economists quietly complained that she wasn't an economist but a political scientist—and a woman).

Ostrom surveyed communities that organized *common-pool resources*—including forests, fishing areas, fur trapping regions, irrigation systems, pastoral grazing commons—over long periods without formal contracts, rules, or governmental enforcement but didn't unravel into market failures such as overuse or underinvestment. She combined her data with theories and findings in experimental and evolutionary game theory to identify a number of features that enable groups to solve their collective action problems. In effect, Ostrom devised a game-theoretical recipe for solving collective action problems in the provision of public goods.[3] It has seven features:

1. Boundary membership rules ensure that parties to the collective are clearly identifiable to one another.

2. Rules allocate benefits proportionately to the contributions made and costs incurred by members of the group.
3. Rules are made and modified by members of the group.
4. The group selects monitors who are accountable to members or are members themselves.
5. There are graduated sanctions for those who violate rules.
6. Conflict resolution methods are quick and low cost.
7. Local group organization is permitted by higher levels of authority.

Such rules can emerge either by explicit agreement or as convention, tradition, or custom by some process of Darwinian cultural selection (we'll explore this process, uncovered by evolutionary game theorists, in chapter 10).

Ostrom was widely lauded in her lifetime for showing how bottom-up, grassroots organizations can provide something to members that individuals, governments, and competitive markets couldn't or wouldn't be able to secure. We can apply Ostrom's insights to understand the persistence of coalitions and collectives that rational choice theory tells us couldn't last due to defection and free riding.[4] This will be of great importance in our examination of how widespread and persistent market failure is in the real economy.

Back to Labor Market Monopsony

Recall the models of the labor market from chapter 7. New classical economic theory treated the labor market as slightly imperfect due to friction but as one that, when left to itself, always remains at or returns to zero involuntary unemployment (or as close to it as possible): that allowing for job search time, any worker who doesn't have a job doesn't want one and any firm with a vacancy it can't fill isn't offering a high enough wage to fill it. It's a competitive market, mostly. Besides, when monopsonies do arise in the labor market, economic theory tells us that, like monopolies, they cannot last. Why not?

The suggestion that monopolies can't last, as we've discussed, is well known in economics. Monopolists make rents—profits that are unearned in the sense that they reduce the allocative efficiency of the market. These rents are signals to others to enter the monopolist's market and undercut them enough to scoop up most of their rents. As more and more competitors enter the market, the monopoly is eventually eroded. Besides, as we've seen, everything is a monopoly and nothing is: no pair of goods is identical (and there are always transaction costs), but every commodity has some substitute. Firms are on a spectrum from perfect competition to monopoly, but none exist at either end. RCT requires that when a market veers too close to monopoly, self-corrective forces take over, mobilizing more rent seeking to put a stop to it. Even when there are just a few firms selling in a market and colluding to fix prices, there are strong incentives for each firm to defect from the coalition. One firm will break the explicit or implicit price-fixing agreement and cut prices just enough to scoop up extra rents at the other firm's expense. Then the scooped firm cuts prices some more, and so on. In the long run there's nothing to worry about. The deadweight loss will be eliminated and the consumer surplus restored. That's the laissez-faire take on monopolies espoused by pure theory.

Even if this process eventually unravels monopolies, will it work equally well for monopsonies, including labor market monopsonies—the ones we care about? Ostrom's work gives us a handle on how and why it won't always or even often happen in the labor market and why a small number of buyers of labor can continue acting as a unified coalition indefinitely.

Recall that in the labor market there are lots of sellers (all of us working stiffs) and few buyers (the firms that employ us). It's not just the big businesses. Think about retail stores along the same street or strip mall hiring salespeople. Each store has a strong incentive to buy labor at the lowest price they can. If the stores can gang up, either formally or informally, to offer the same wages, they're acting as a monopsonistic coalition. Will it hold up?

Ostrom's research shows that the answer can be *yes*. Once the few buyers of labor in a market of many sellers of labor figure out a wage rate that secures them rents, it will be relatively easy for them to maintain their coalition. That's because they will be in the same position as Ostrom's groups maintaining a common resource pool—areas to catch fish or trap fur—with enough incentives to continue to do it forever and few temptations to defect from the coalition (i.e., to catch more fish—better workers—using slightly better bait—a little more money).

It's not very difficult for firms buying labor in the same market to satisfy all of Ostrom's conditions for solving their collective action problem. I've quoted Smith on this point, about how shop talk among individuals almost inevitably results in agreements to raise prices, twice already. It's even easier for employers to contrive to lower wages and keep them low. Remember what the lunch conversation at all those Rotary Clubs, Lions Clubs, and local trade association meetings ends up being about.

Let's go over Ostrom's conditions for solving collective action problems with an eye to how easily employers can satisfy them and how much harder it is for workers to do so.

1. Clear group boundaries. In much retail trade, geographic boundaries will suffice to isolate a small group of buyers of labor. The location of employers in a region and the obstacles to relocating a small business, such as a retailer or a supply-chain enterprise, draw geographic boundaries around the group. In other markets, including capital-intensive but geographically distributed ones, the boundaries around a group of employers will instead consist of the capital costs of entry into the market in which they sell, the regulatory environment in which they operate, and the political environment a small number of large businesses can create by lobbying. The boundary makes it to easy identify those who are members of the coalition and those who are not and creates difficulties for nonmembers trying to access the common pool. Enterprises that have built large, capital-intensive infrastructure or already own primary product inputs (local mines, for example) are difficult to move but are themselves protected from invasion by competitors.

    The employer's common-pool resources—the workers they need—can't easily move, either. They can't uproot themselves instantly, and most don't want to. That's what makes them a common-pool resource for the employers.

2. For a coalition of common-pool users to be stable, the rules governing the extraction of the resource—fish, beavers, water, or workers—have to be suited to local needs and conditions. The costs of extraction and benefits must be proportional for each participant and flexible enough to survive changing circumstances. Business owners and hiring managers are in the best position to know what the relevant circumstances are (their local needs and conditions and those of their competitors). They will know their own marginal productivity of labor—how wage rates affect their total costs of production. To the extent their markets are similar, they can assume other buyers of labor face the same costs and are similarly motivated. The rules to which they adhere need not and indeed often cannot be explicitly stated; they often emerge through a process of experimentation, imitation, and correction. Wage rates offered for

similar jobs are easy to vary until they reach a level that is more or less optimal for coalition members, and that level can harden into a locally enforced implicit convention. In some local labor markets, a practice of not poaching workers can easily become a norm and then an unspoken rule. In other markets the requirement of noncompete clauses may become imposed uniformly by employers, even in low-skill industries such as fast food and without explicit agreement among the buyers of labor to impose them.

3. Those affected by the rules participate in modifying the rules. Because the employers all have approximately the same access to the same information relevant to production, they are in a position to recognize rule modifications that would be mutually advantageous. Among a small number of employers, informal networks of communication enable coordinated implementation of such changes.

4. Community members can monitor other members' behavior. Where labor is the costliest input of production or the most significant variable cost in the short and medium terms, employers will be strongly motivated to monitor their own and other firms' "job separations"—layoffs and firings—and the reasons for them. This will be especially true in small businesses and ones with very high wage rates, where tracking the decisions of individual buyers of labor to change their personnel is relatively easy.

5. Violations of the coalition's rules are punished by graduated sanctions. Here the social network of employers will be effective in enforcing coalition-maintaining behavior. When one member of the coalition of employers breaks ranks by offering higher wages or other benefits, the other members can easily discipline the free rider. They can gradually drop the prices they charge their buyers until the free rider returns to cooperative behavior. The social network of managers facilitates the coordination, communication, and graduation of any sanctions.

6. Disputes among coalition members can be resolved at low cost. Many of the factors already mentioned make dispute resolution cheap in these groups. Since the coalition is small and well informed about the inputs that members buy and the consumer markets they sell into, members of the group can easily resolve disagreements on facts relevant to production decisions. Insofar as their motives are similar, disputes should resolve themselves quickly.

7. The rule-making rights of community members are respected by outside authorities. This is a relatively easy condition for a coalition of employers to arrange. They do not have to harness outside authorities to enforce their rules either on themselves or on the sellers of labor. They only require the freedom to enforce their own rules. "Right to work" laws are a prime example of the sort of non-interference granted by outside authorities. As we'll see, they contribute to the persistence of the resource—unorganized workers—the coalition of employers can continue to extract.

Compare the ability of workers, employees, job-seekers or job-holders to solve their collective action problems. If it was possible in the past, it's practically impossible today. It has always been true that there are many sellers of labor and far fewer buyers of it in any market. That's already a good start to explaining why workers can't solve their collective action problem. It's obviously much harder for large numbers of individuals facing a collective action problem to coordinate. There are too many of them to monitor each individual's compliance with any agreement they hope to make among themselves.

There are two other reasons that together make the problem of collective action much harder for the sellers of labor to solve. Thomas Schelling nicely illustrated one of the problems faced by sellers of labor with the case of National Hockey League hockey players and the helmets none of them wished to wear. If only one player wears a helmet, he will reduce his visibility and be susceptible to ridicule by fans and other players. If all players wear helmets, they all pay the same costs in visibility and reputation and secure the benefit of reduced head injuries. If all wear helmets except for one, the single bareheaded player will secure a competitive advantage. The rational hockey player will therefore forgo a helmet regardless of what others do. As a result, few players in the NHL used to wear hockey helmets. Hockey team owners didn't have any incentive to make players wear helmets. Owners knew that fans come to NHL hockey games to see fights and that helmets reduce fighting (who wants to hit a hard helmet?) and so reduce profits. But if all players wear helmets there will be no relative differences in performance due to helmets and all players will be better off. The solution to the problem facing each hockey player is to participate in a player's association (a labor union) that can enforce helmet wearing in the league.

But if the team owners face reduced revenue from an overall decline in player performance or less of the fighting on ice fans come for, they'll want to discourage this collective action solution to the individual players' problem. They'll reward individual players for not wearing helmets. Now combine the transaction costs problem of unionizing individualistic athletes with another problem recognized by a very perceptive economist, Robert Frank. We'll see how they combine to put workers at the mercy of employers.

Team owners, besides opposing unions, may be glad to accept a league-imposed total salary cap (many professional leagues do), in which case the rewards to players will be relative. That is, players will find themselves in a zero-sum game: more money for one player means less for another. It's a situation that may increase the relative cost of wearing a helmet and the relative benefit of not wearing one. But it's not just professional athletes who face relative reward problems. Almost all of us do, which makes the wage earner's collective action problem much harder to solve.

In most cases, the most important and certainly the most expensive consumption good a worker wishes to secure is a home. There are many reasons why an individual supplier of labor wants to acquire the most expensive residence they can afford, including access to the amenities accompanying it, including many positive externalities such as higher-quality schools for their children. But the rental and real estate markets are characterized by auction-like conditions in which individuals bid against one another for a limited supply of consumption goods. Thus each bidder is motivated to have not the largest amount of wealth or income in absolute terms but the highest amount relative to other bidders in order to outbid them. Sellers of labor face a serious problem of earning relatively larger rewards in competition with all other suppliers of labor. This results in the incentive to free ride or defect from any coalition of employees if what they need most is a high enough wage to outbid other buyers for consumption goods.

The relative reward problem faced by suppliers of labor makes joining a union a nonstarter for the rational worker. If other suppliers of labor pay for the provision of a union or withhold their labor in strikes, then the supplier who does not do so earns more and can outbid other suppliers for the most important consumption good they all seek. If the sellers of labor in one market pay for a union and agree to withhold their labor from that market, they will also be unable to bid against workers supplying labor in other markets to secure their most desired consumption good. This will almost certainly occur when the buyers of labor can raise the sellers' costs of solving collective action problems by, for example, securing the passage of "right to work" laws. These laws, in force in a majority of states in the United States, ensure the right of workers to decline to join labor unions while requiring that they receive all benefits labor unions provide to their paying members. They are more aptly called "right to work at lower wage" laws, because they provide incentives to sellers of labor to free ride and thus unravel any coalition of wage earners. Combine such laws with the importance of relative rewards and the obstacles to effective collective action become much greater for workers than for employers.

In the distant past—the middle of the twentieth century—some wage earners were solving their collective action problem by joining labor unions that confronted the monopsony of buyers with a monopoly of sellers. This was especially true in the US auto industry, where there was a monopsonistic coalition of three automakers employing workers and a single labor union monopoly representing them—the United Auto Workers. Between them, there were few strikes but a number of successfully negotiated wage settlements. Why?

The monopsony coalition of automakers and monopoly of autoworkers faced a set of strategic action problems that game theorists could model. Here is a relatively simple game-theoretical analysis of a labor union and management strategic interaction problem:

	Union	
	Don't strike	Strike
Management Raise wage	II 2d 2d	III 1st 4th
Don't raise wage	I 4th 1st	IV 3d 3d

Union preference ranking: III > II > IV > I
Management preference ranking: I > II > IV > III

Game theorists call strategic interactions with payoffs of this ordinal structure *stag hunts* after a fable told by Jean-Jacques Rousseau in *The Social Contract*. In Rousseau's fable two hunters could successfully hunt stag, with a high payoff for both, only if they worked together. Separately, each could successfully snag a hare, but if one hunted hare and the other stag, the solo stag hunter would get nothing. The highest attainable payoff to each is when both hunt stag, but each hunter needs to be confident the other sees this, too. Otherwise each hunter risks coming up with nothing while the other gets at least a hare. It was Schelling, again, who suggested that when the cooperative payoff is highly salient to both parties, when it is, in his words, a "focal point," it's worthwhile for each party to take a risk to go for it. In negotiations between a labor union and management, an arbitrator can draw both parties' attention to the focal point, a win-win outcome. (Not for nothing did Schelling win a Nobel.)

If the negotiations between the UAW and the Big Three automakers (Ford, GM, and Chrysler) in the 1950s had this structure, it would be no surprise there were few strikes and lots of pay increases. But over the last fifty years or so, the strategic interaction payoffs to management and unionized workers in many industries have been changed radically, almost always in ways favorable to management. In many industries technological changes have enabled firms to shift production quickly from region to region with low transaction costs. In other cases the freeing up of financial markets has allowed companies to invest in new factories in countries where unions don't exist. This has improved the employer bargaining position in labor-management disputes since they can credibly threaten to move production to regions with lower wages. Many firms have used this threat. It used to be hard to move factories, and too risky to build them in other countries, but not anymore. These changes have radically reduced unions' bargaining power by making strikes ineffective.

Of equal importance is firms' use of the rents their market power secures to lobby governments. In the United States, business lobbies have encouraged more than half of the states to harken to the call of "freeing" their workers from coercive trade unions. After all, unions spend members' money to elect politicians that management doesn't like. Maybe some workers don't like these politicians, either. In most states, "right to work" laws freed workers from the obligation in a unionized workplace to join the union and pay dues. With such laws in place, workers could benefit from union gains without paying union dues, turning the stag hunt game between labor and management into a public goods prisoner's dilemma among workers. An across-the-board wage increase is non-rivalrous and non-excludable among everyone in the same job category. But if Abe doesn't have to join or pay dues to the union that negotiated the deal, he'd be irrational to do so. Betty reasons the same way, as does Charlie, Dawn, Ed . . . The result is the end of the union altogether. All in the name of "freedom."

	Enough other workers	
	Join union	Don't join union
Alf Join union	II Good Good	III Bad Worst
Don't join union	I Good Best	IV Bad Bad

The preference rankings are

Alf: I > II > IV > III
Enough others: I ~ II > IV ~ III

All the workers go for box I and they all end up in box IV, where management wants them. We've seen this outcome before in free riders unraveling the provision of a public good.

In short, one of the two parties found a way to change the stag hunt game to another one that would achieve their more preferred outcome. The combination of "right to work" laws and greater flexibility in where firms do their manufacturing enabled a monopsonistic coalition of employers to secure rents by eliminating unions altogether across what came to be called the Rust Belt.

New classical economic theorists' commitment to market-clearing equilibrium blinds them to this outcome—persistent monopsony in labor markets. The only kind of nonoptimality in labor markets they can recognize is the friction of job searching. If your only tool is a hammer . . .

But when the market failures begin to pile up, you need a new tool to understand what is going on. To understand the real economy's market failures, and especially if we are going to fix them, we need game theory. That's because the real economy is never near enough to the perfectly competitive general equilibrium in which we are all price takers. It's nothing but market failure—everywhere.

9 Fated to Fail: The Financial Markets

A financial market has three types of agents: two principals and a middleman. The buyer is the party that needs someone else's money and buys it in the form of loans they borrow or stocks and bonds they sell. The seller is the party with money to lend or invest in bonds and stocks. The middlemen, or market makers, are the stockbrokers, banks, hedge funds, and insurance companies that make deals, bringing buyers and sellers together. In a modern developed economy, there are large numbers of sellers of cash to invest, mainly us jobholders who regularly put aside smallish quantities for our retirement. There are a smaller number of companies—manufacturers, builders, agribusinesses, health care service providers, hardware and software tech businesses, and so on—that need to borrow or sell shares to stay afloat till their projects start generating income. And there are an even smaller number of market makers—investment banks, mutual funds, insurance companies, and hedge funds—that connect the large number of smallish investors with the smaller number of big capital-consuming firms.

Financial markets are the most important markets for all of us, whether we own stocks or not. It's not obvious why they are so important. It's even less obvious why they don't work—can't work—the way economic theory says markets should. Economic theorists pretty much accept that the financial market is the domain of the greatest market failure. But they're at a loss for what to do about it. The explanation of the problem is all game theory—the theory of strategic interaction. That doesn't make the problem soluble, but at least it shows why it isn't. By explaining it in the most general terms, using the resources of game theory, we'll show how serious the problem is.

The problem starts with what Adam Smith thought his real insight was—not the competitive market but the division of labor. The invisible hand was not the big original idea of *The Wealth of Nations*; Bernard Mandeville had already thought of it. Competitive markets make for economic growth, according to Smith, largely because they harness the division of labor to increase productivity—making more consumer goods with fewer production inputs, or making consumer goods faster or more cheaply or better, continually. The first three chapters of Smith's book were given over to showing how and why the division of labor emerges and what circumstances encourage its emergence. The next chapter was devoted to the origin and use of money, since the division of labor cannot emerge without it. The reasons are obvious: even if you make nails more cheaply, better, and faster than anyone else, you can't eat them while you make them; you have to buy what you need to survive while making the nails that people will want when you've finished them.

Money Is a Claim on Future Consumption

Recall our discussion of money as a claim on future consumption back in chapter 5. We need to add some detail to that discussion in order to illuminate the crucial job a financial market has in an economy.

Money is required in any economy in which making things that people really want takes more than a few days. It solves this problem because, along with its other features, a unit of money is a claim on future consumption. People who make more of anything than they immediately consume can postpone their consumption of it, selling their surplus for money that they can use later if the money is accepted as payment in the future.

Without a device that works as a claim on future consumption, no economy can grow through the division of labor, nor can it grow through the production of goods it takes time for specialized labor to make. What kinds of goods are these? Agricultural crops, energy sources (from split logs to solar power arrays), clothes, shoes, pots and pans, buildings, roads, vehicles—in other words, capital goods, or things you need to make other things. Growing wheat for flour to make into bread, raising sheep for wool to spin and weave, and building a road or canal or railroad track all take time, sometimes years. The designers and workers can't fend for themselves in the meantime. Where can they get what they need to survive till their projects are finished and sold?

If the producers of immediate consumption goods make more than they can consume, they can provide some of their excess goods to those engaged in longer-term production activities while they are completing these projects. But the producers of consumer goods have to rely on promises to return more of the same (or other) consumption goods to them later, when the longer-term project is finished and can be sold.

I'm belaboring the obvious here to home in on this function of money as indispensable for the production of capital goods, which often take a long time to imagine and create, use a lot of inputs and specialized skills to make, but don't wear out or get used up quickly once produced. Combine money's function as a claim on future consumption with its ability to solve the problem of the double coincidence of wants and you have a social device, an institution, that makes possible long-term productivity increases and therefore economic growth.

Does this require economists to treat money as a real, causal factor in the economy? Think back to chapter 5 again. We saw that the microeconomics of rational choice theory's buyers and sellers makes money a mere convenience without a real role to play. In an economy of relative prices no one cares about the nominal units. Money is, as classical economists and new classical macroeconomists say, neutral: changes in money prices are treated by rational agents as only a corrigible indicator of changes in real prices. The function of money as a claim on future consumption doesn't force these economists to change their views about restricting its role to accounting. The need created by economies in which capital goods are produced simply complicates the double coincidence of wants problem to include trade between goods that are currently available and goods available only in the future.

Economists like John Maynard Keynes based their theories about money on assumptions about radical uncertainty and our departures from RCT rationality. They made much of money's role as a claim on future consumption to give it a causal role in the economy. People who have more or make more than they can immediately consume can gamble with it by using it to place bets in the stock and bond markets as well as the bingo parlor. In a market characterized by some or a lot of radical uncertainty, where there are no well-behaved probabilities to calculate, some people will take huge risks and others the least risk possible. Over time their "animal spirits"—for example, preferences for loss aversion and willingness to discount hyperbolically—will drive markets away from allocatively efficient equilibriums and keep them there. Money will have a causal role in the economy, generally one that produces departures from optimal outcomes.

But even if these economists are wrong about the causal role of money in the economy, the problem of producing capital goods that money solves spawns another, overwhelmingly consequential, market failure problem for every economy.

What Financial Markets Are Supposed to Do

I described financial markets briefly at the beginning of this chapter. But let's think about them through the lens of game theory. We'll see more clearly how they generate collective action problems due to the transaction costs they impose. To do this we'll need the labor market insights of Mancur Olson and Elinor Ostrom from chapter 8.

The makers of capital goods—production goods—often need huge amounts of consumption goods and other production inputs. They need to find the producers of consumption goods who have surpluses they can spare and are willing to part with for later repayment (with interest). Consumption goods producers with surplus have the problem of finding the producers of capital goods who will buy their excess.

But there is a serious asymmetry between buyers and sellers in this market. The producers of capital goods (think Boeing) almost always need *huge* quantities of consumer goods delivered over long periods of time while producing their capital goods before they can sell them or put them to work. Almost none of the many owners of surplus consumption goods (think small farmers) have huge supplies of them—not enough, by themselves, for what any one capital goods producer needs over even a small part of its production process.

The vast number of small consumption goods producers need a mechanism for combining their smallish quantities of surplus consumer goods into big bundles and getting them to capital good producers. And the capital good producers who need large amounts of money to pay for the huge amounts of consumption goods they buy need to figure out how to later pay it back to all those lenders of small amounts of money. There is also an asymmetry in numbers between these two types of producers. There is usually a far smaller number of capital goods producers than consumer goods producers. How will the smaller number of capital goods producers find the many small consumer goods producers they need to trade with?

What these two types of parties need is a market that provides buyers with all the immediate consumption goods they need and sellers with claims on future consumption, not just of what they immediately produce but anything and everything they might want to consume in the future. They need an institution that solves the problem of their double coincidence of current versus future wants. They need financial markets. And it's not just these two parties who need such markets; the rest of us do, too, if we hope to have better, longer, healthier, and happier lives, and if we want to enjoy the fruits of economic growth via the division of labor. Efficient financial markets solve the most important problem every growing economy eventually faces.

A financial market solves the problem of transaction costs that the buyers and sellers of postponable consumption face: combining all the different consumption goods that large numbers of small producers can postpone consuming themselves, finding the capital goods producers who need different amounts of these consumption goods, and then solving the double coincidence of wants problem that all these parties face. A competitive financial market reduces these transaction costs to their minimum: deal makers compete to put buyers and sellers together to complete transactions. The competition between deal makers improves the division of labor in the financial markets in the form of new financial instruments that reduce the transaction costs for buyers and sellers.

Middlemen on the financial market buy all the goods that are surplus to their owners' immediate consumption and combine them into bundles they sell to capital goods producers. That's where money comes into the picture. When us little guys save or invest we are, in effect, selling claims on immediate consumption goods to a bank, broker, or retirement fund in exchange for claims on future consumption. When the capital goods producer borrows money or sells stock, they use the money to buy current consumption goods needed immediately. The money they pay out is a promise to provide future consumption goods. Financial institutions—banks, brokers, hedge funds, mutual funds, and insurance companies—solve transaction cost problems. They provide the market that brings together buyers of immediate consumption and sellers of future consumption by buying lots of small quantities of postponable consumption, which they package into large bundles they sell (or lend) in exchange for promises to provide larger quantities of future consumption. Then they divide and distribute the claims on future consumption among the large number of small sellers of today's postponable consumption.

A perfectly competitive financial market is probably the single best thing that can happen to an economy. A perfectly competitive market, we know already, is allocatively efficient. It puts every input of production to its most efficient use for making the biggest bundle of consumption goods that people actually want. When the economy needs capital goods to do that, the financial market is indispensable. It's the market that makes the biggest difference to the whole economy's allocative efficiency over time because the quantities of goods traded on these markets are the largest, at least in their money values, and their impact on the productivity of the whole economy the greatest.

At its simplest, the perfectly competitive financial market allocates inputs by means of the interest rates that capital goods producers are willing to pay to borrow postponable consumption goods. The interest rate is the price demanded by sellers for their postponed consumption and offered by buyers to secure immediate consumption of the goods they lack. In a perfectly competitive market the "law of one price" will prevail, everyone will be a price taker, and interest rates will ration postponable consumption to its most productive uses—the processes with the highest marginal productivity in creating consumer goods. The interest rate will "decide" who gets to produce which capital goods, when, and where, by one producer's willingness to pay a higher interest rate than others. Every firm's return on its investment will be equal to the interest it pays. There will be no rents in the market for loans any more than there are profits in other markets of the perfectly competitive economy.

Firms also secure productive capital by selling stock, issuing bonds, bringing in investing partners as new owners, and more. Economic theory has shown that when risks are measured accurately under conditions of perfect competition among rational agents, the returns to bonds, shares, and partnerships, in terms of future consumption, will be exactly the same. Economists call this the "law of one price." So we can use the interest rate as a stand-in for the dividend rate and other vehicles for rewarding those who postpone consumption.

Now it should be evident that the closer real financial markets come to the ones in a perfectly competitive economy, the better it is for the real economy in which they operate. The trouble is that real-world financial markets are even more likely to be riddled with market failures than labor markets. Financial markets are the ones where we most urgently need game theory to help identify, mitigate, and prevent the rent-seeking and price-setting behaviors that are most devastating to economies—developed, developing, and undeveloped. And the biggest challenge we face in using game theory to design and improve institutions to prevent this is that the job can never be finished.

Financial Markets: Where Market Failure Runs Wild

Recall that a financial market is characterized by three different kinds of agents. First, a large number of individual sellers of small amounts of postponable consumption: working people who save some of what they earn. Second, a smaller number of firms (usually operated in the interests of their managers, who are not their shareholders) making capital goods. Third, an even smaller number of market makers: banks, brokers, and others who solve the transaction cost problems of trade between the first two groups.

As we have seen, members of the first group, the largest one, have to pool what they don't immediately consume. This is a collective action problem that they can't efficiently solve by coordinating among themselves. There are too many of them, with many different things to sell in many different amounts; their interests are diverse and their levels of trustworthiness vary. They are unlikely ever to form coalitions that monopolize their markets.

The second group, producers of capital goods, is smaller in number than the group producing immediate consumption goods. But its size is an advantage; there are few enough producers, especially of the most expensive, complex, and durable capital goods, that they can occasionally solve some of the collective action problems they face. If they establish monopolistic coalitions (think trade associations), these are likely to be stable, too. They can harness the rules Ostrom identified that enable groups to provide themselves collective goods. Such firms are also sometimes in a position to form monopsonistic coalitions in their purchases.

The third group in the financial market, the middlemen—financial institutions—is the smallest in number. Their members are more similar to one another in many ways than are the members of the other two groups: they have the same interests (making rents), roughly equal knowledge about buyers and sellers, a good understanding of the rules that the market obeys and who enforces them, a sense of how effectively the rules are enforced, and an appreciation of the benefits and costs of solving collective action problems they may face.

Because they are numerically the smallest group, market makers have the fewest problems forging coalitions among themselves to solve their own collective action problems. Their numbers are not only relatively smaller, but small even in absolute terms. In some nations the number of banks can be counted on the fingers of one hand. So market makers can establish coalitions that monopolize the sale of postponable consumption goods—loans to capital goods producers. They can also organize the same or different coalitions that monopsonize the purchase of postponable consumption from sellers. The ability of market makers to monopsonize is probably much greater than their ability to monopolize. The sellers of postponable consumption that banks buy from are each small in size, but the group of sellers is very large in number. These sellers can't easily organize to bargain with the market makers. The monopoly loan-selling coalitions often provide loans to only a small number of makers of very costly capital goods, so market makers' ability to secure rents on this side of the exchange may be more limited.

Since there are huge quantities of money at stake for both buyers and sellers, market makers negotiate for both. They charge a "transaction cost" to put buyers and sellers together. It's a small percentage of the prices negotiated, while the rest of the money proceeds to the members of the monopoly/monopsony coalitions who need it immediately. That's where the opportunity to secure substantial rents for the market makers comes from. What market makers buy and sell is money, credit, or means of payment: a homogeneous, infinitely divisible commodity with almost no cost of storage, depreciation, or transfer. Given the large numbers of trades these coalitions engage in, they have less trouble than other monopolists or monopsonists experimenting to identify the optimum pricing strategies from day to day and market to market. All are roughly equally well informed about the costs and benefits of building and maintaining coalitions. They also know the government regulations currently in force to limit or mitigate market failure produced by such coalitions (as monopolies or monopsonies), and they have strong incentives to flout or circumvent them.

Perhaps the most significant departure from the conditions of perfect competition is that financial markets fail to even come close to fulfilling the requirement of full information. In the perfectly competitive market everyone is on a level playing field when it comes to information relevant to maximizing their preferences. Because prices internalize and transmit information continuously and instantaneously, not only does everyone have the same information but they have all the information they need to produce the allocatively optimal outcome for themselves and thus, via the invisible hand, the whole economy.

In the real world of financial market makers, none of the three kinds of participants are in the same informational position. The market makers are in the best position of all when it comes to making choices that will secure rents. Capital goods producers know the most about their costs and market prospects and have some incentive to misrepresent their beliefs to one another and the financial market makers. The size of the loans the capital goods producers need to buy from the market-making monopoly loan sellers is large; therefore the middlemen selling loans have strong incentives to inform themselves about the capital goods firms in order to accurately calculate risk and their level of information will approach that of the large firms that buy loans from them.

The large number of small sellers of postponable consumption know much less than members of the other two groups. They'll know almost nothing relevant to making optimal decisions about what they should sell, how much to sell, whom to sell to, and at what price to sell. And the costs of informing themselves are great. The playing field is tilted steeply in favor of the monopsonistic buyers of what small sellers have to sell, just as in the labor market.

What's the predicable result of the differences between members of these three kinds of groups in their opportunities to be market-failure-producing, rent-seeking price setters? It will be significant departures from allocative efficiency in the most consequential market in the whole economy. And it will be the market makers, the financial institutions, that do the most damage by securing the most rents.

This is where the study of strategic interaction comes into the picture for market-making firms, governments regulating financial markets, and game theorists consulted by both parties in order to maximize rents or suppress them.

Coalitions of market makers and aggregating buyers will repeatedly find themselves in strategic interactions that require both parties to adopt the tools of game theory to maximize their rents. Recall the cooperative stag hunt game from chapter 8. Rent seeking can start with market makers colluding in trades with both small sellers (easier, due to their large numbers and small sizes) and large buyers (harder, since these buyers are powerful, small in number, and highly incentivized to secure full information). Coalitions of market-making firms will also have the resources to lobby legislatures, suborn regulatory bodies, evade enforcement or minimize penalties, and convince governments that they are "too big to fail."

Globalization has greatly increased many of these market-failure-producing opportunities for rent seeking. International trade organizations such as the World Trade Organization have required that national financial markets be opened to foreign buyers and sellers of postponable consumption. Where trading approaches perfect competition, this should in principle enhance the allocative efficiency of the whole world's economy. However, fifty years of history have shown that the movement of large amounts of money in and out of small developing economies has been very harmful. Opportunities to chase currency arbitrage, along with money laundering and the competition between jurisdictions for the international finance business, has further incentivized the formation and maintenance of monopoly and monopsony coalitions of market makers. Not all of their strategies are openly acknowledged, visible to the public or regulators, or even legal.

It's in the strategic interaction between market-making firms and governmental regulators that the importance of game theory becomes most evident. This is the domain of what game theorists call *arms races*.

The term comes from the original domain of game theory—the competition between the United States and the USSR during the Cold War. Over forty years each side repeatedly developed new weapons to counter the other side's weapons innovations. First aboveground missiles were vulnerable to attack, so they were succeeded by underground missiles in silos. When these became vulnerable both parties moved to launching missiles from submarines. The result was the development of antimissile defenses. These provoked multiple independently targeted warheads, which eventually spurred research on the US Strategic Defense Initiative. That seemed to be the last move in this arms race before the process of nuclear disarmament began. The nuclear arms race ended, at least temporarily, in the 1990s.

The twentieth- and twenty-first-century history of the regulation of financial markets reflects a similar arms race, one we should expect to continue until the ingenuity of participants in these markets is exhausted. The rent seeking of market-making financial institutions in the nineteenth and early twentieth centuries resulted in the creation of the Federal Reserve Bank and the establishment of its oversight of national banks after 1914. The rent seeking of unregulated regional banks before the Great Depression, which began in 1929, led to widespread bank failures in 1932, which resulted in the Banking Act of 1933 (also known as the Glass-Steagall Act), designed to curb banks' rent seeking. Market makers still found opportunities for rent seeking within this set of regulations. That culminated in the subprime mortgage meltdown of 2008. By 2010 a new set of regulations was put in place—the Dodd-Frank Act. Its regulations only pose a new set of obstacles that rent-seeking firms and individuals will find ways to circumvent. One side employs game theory to design and improve institutions to reduce or eliminate at least some kinds of rent seeking. The other side uses game theory to find opportunities in the new regulations to seek rent.

So long as the financial market is the only solution to the problem of allocating postponable consumption for economic growth, the strategic interaction problem will be with us. The disparities between the numbers of buyers, sellers, and middlemen, along with the differences in their transaction costs of doing business, will preserve the rent-seeking opportunities that make long-term economic growth suboptimal. All we can do is elect governments that will search for regulatory second bests, knowing full well that the regulated will find workarounds.[1]

The problem will always be with us, and it will get worse.

The Rich Will Always Be with Us

The New Testament quotes Jesus saying "the poor you will always have with you," and the eighteenth-century economist Reverend Thomas Malthus tried to prove it.[2] Maybe he did, but the rich will always be with us as well, and they will be getting richer and richer.[3] What's more, their wealth cannot help deforming the market to bend further away from perfect competition (if it was ever near perfection to begin with). The irony of Smith's prescription in *The Wealth of Nations* is that the market's efficient operation unravels itself, producing ever-increasing inefficiency and more and more market failure resulting from greater inequalities of wealth. The economic institution that does all this work, moving us away from perfect competition, is the financial market, and there seems no way to replace it with something better. That's why capitalism is the worst economic system—except for all the others.

Here's the problem in its hardest nutshell. The perfectly competitive market directs productive inputs to their allocatively most efficient uses, and this includes labor inputs. Sellers of labor will be channeled by the market into jobs that maximize their productivity. Individual sellers of labor have different ambitions and abilities, which will determine their productivity. Therefore incentives are required to get individuals with different levels of ability and ambition to do the work best suited to each. If wages reflect productivity, then the individuals who are more productive, due to their ability and ambition, will have to be paid higher wages than those with lower productivity. No rents will be earned at the start. Wage inequality will simply reflect inequality in productivity between workers resulting from their differences in skill and different preferences for leisure.

These wage differences produce inequalities in income and then wealth. Wealth is the amount of postponable consumption goods you have. If financial markets exist, these postponable consumption goods can be sold for claims on future consumption, at interest or some other promise of receiving more later in return for the amount sold now. The result is a further increase in the amount of postponable consumption at the disposal of those whose labor has higher productivity. This can go on forever. Those who work harder and better make more stuff and can afford to postpone more of their consumption. The result is that over time the perfectly competitive market produces ever-increasing inequalities of wealth among wage earners. Then the market invests that wealth at interest, which increases the inequalities of wealth. How long before the wealth differences begin to be reflected in price-setting market failures?

Market failure is inevitable. The wealthy must work with the financial markets to invest their postponed consumption efficiently. Coalitions of rent-seeking financial market makers, as monopolists or monopsonists, will bargain with the wealthy, sharing rents with them while sending most of their postponed consumption to the most allocatively efficient capital uses. The result will be greater wealth disparities and another cycle of increase.

The process will continually tilt the playing field away from allocative efficiency for the whole economy.

There are other ways that inequalities of wealth will iteratively deform an economy. What starts out as an economy that might look perfectly competitive—if you squint—increasingly becomes one in which wealth inequalities generate sufficient surpluses to turn more and more participants into price setters. Their exercise of market power may emerge in any number of ways. As buyers of the labor of others they may secure monopsony rents. As producers of capital goods they may exploit asymmetries of information in the financial markets or form coalitions to set prices for their products. As discoverers and inventors of new technologies, their trade secrets (or patents, if they can organize the institution of the patent) will put them in a position to enforce monopoly pricing. Financial market makers will have incentives to join coalitions with those who amass increasing amounts of wealth in the form of claims on future consumption. The only parties to the economy who will be unable to move from price taking to price setting in their trades will be those unable to produce enough consumption goods to have significant surpluses. These will always be the lower income, less wealthy participants in the economy.

However, the rents that begin to be secured, moving the market further from allocative efficiency, will also drive economic growth. If prices on financial markets send accurate signals about the most productive directions in which to send postponed consumption, the economy will be kept on a trajectory approaching optimal growth. This trajectory will be maintained even if the financial markets work imperfectly. Those with substantial wealth will have an incentive to send resources to the long-term production projects with the highest return to the economy as a whole, even as their interests are to secure continuing and increasing rents from these future returns.

Add to this process the multigenerational impact on market power resulting from inheriting wealth. Inheritances beget increases in market power and carry inequalities beyond the lifetime of any individual agent. Those with wealth to invest want to ensure and maximize the continuing flow of rents to their descendants. They are prepared to invest in long-term production processes whose returns will not arrive before their deaths. To the extent financial markets allocate inputs efficiently, this proclivity to amass wealth and pass it on to later generations will also contribute to keeping the whole economy on a trajectory that has a chance of approaching optimal growth.

Alas, so organized, the inequalities in wealth that are produced, and their consequent increases in market power, price-setting opportunities, and rent-seeking successes, must drive the economy further and further away from an allocatively efficient outcome. It's easy to tilt a market away from perfectly competitive to one rife with market failure.

The cure is obvious. But it's worse than the disease. All we need to do is re-level the playing field: once every century, or every generation, or every decade, eradicate all inequalities of wealth. Redistributing the proceeds of a confiscatory death duty or a swinging tax on amassed wealth would make all participants in the economy equally well endowed. It would, at a stroke, eliminate the sources of market power: the rents deforming the market away from allocative efficiency. Now, given the distribution of abilities and ambitions, we must allow the whole process to start again, knowing that a decade or century hence we'll have to redistribute assets once more. Meanwhile, we'll allow those with postponable consumption to again invest their wealth in the optimal growth of the whole economy.

Alas, this won't work.

What will rational agents do if they know their wealth will be confiscated at some future point? Well, they will almost certainly not make investments in long-term projects whose payoffs will come after the confiscation date. In fact the rational agent will engage in no investment with a payoff that cannot be fully enjoyed before death. No one would be able to leave their wealth to heirs. Though the rational agent may give it to others, it will be subject to the same confiscation in their hands. Most rational holders of wealth will engage in consumption, not savings—purchasing increasingly big-ticket items (think oligarchs' yachts) since they know that whatever goes unspent will be taken from them at regular intervals.

Regular re-leveling of the playing field would, at a minimum, destroy any incentive that the wealthy have to continue to postpone consumption. They would cease to invest it in projects that increase economic growth and add to everyone's well-being even as they secure further rents for themselves. Knowing that one's wealth will be confiscated would incentivize rational agents to consume more and postpone consumption less. Instead of saving and investing in the projects with the greatest return to the economy, the wealthy would fritter away their postponable consumption in economically unproductive activities while reducing the economy's investable resources. Leveling the playing field this way would make irrational many of the economic activities that have prospects of securing rents, including innovations, inventions, discoveries, and other steps that enhance the ability of the division of labor to exploit productivity increases. In the long run, economic growth would grind to a complete halt, resulting in stagnation and eventually the depreciation to zero of all capital stock—the economy's tools of production. At that point the economy may well be distributing outputs efficiently given its inputs. But it would have no resources to invest in capital goods to produce economic growth. If the disease of capitalism is economically inefficient rents, then the cure would be worse than the disease.

Is the disease in fact incurable? Or could we—our governments—design institutions that reconcile the need to keep the playing field level with the benefits to all of productive investment of rents?

There is a strategic interaction problem here faced by two parties: the wealthy and the rest of us, represented by democratic governments. Each side lacks knowledge about the other party's available strategies, the payoff structure of the game, and how many iterations or turns it has. Both sides have to experiment with alternative strategies. For example, if the government graduates the tax rate and experiments with various levels of estate tax over time, the willingness of the very rich to continue to invest in productive opportunities will vary as well. Finding a local optimum may be feasible. But fixing this rate will provide the wealthy, and the market makers in financial markets, with incentives to game the system—to hide wealth or underreport it, convert it to less liquid forms that are harder to valuate, make fake philanthropic donations, set up trust funds for kids, or engage in other evasions.

The arms race between those with wealth and those charged with moving the economy toward allocative efficiency and growth is persistent and never-ending. The strategies of both sides call for more game theory, as does the strategy of economists and others seeking to understand the workings of the real economy. It's all game theory.

10 Economic Theory as Institution Design

Hobbes's Foole and Hume's Knave

In *Leviathan*, written in 1651, Thomas Hobbes identified a problem for political philosophy that he was unable to solve: "The Foole hath sayd in his heart, there is no such thing as Justice; and sometimes also with his tongue; seriously alleaging, that every mans conservation, and contentment, being committed to his own care, there could be no reason, why every man might not do what he thought conduced thereunto; and therefore also to make, or not make; keep, or not keep Covenants, was not against Reason, when it conduced to ones benefit."[1] In archaic English, Hobbes wrestled with the problem of convincing someone not to break the law when it is in his immediate interest to do so. Even when it is unjust why should the foole not do what he wants, as long as he can get away with it? The foole does not deny there are laws that, as a matter of justice, should be obeyed. But he wonders whether acting unjustly is not sometimes in one's interest and thus to be recommended by reason, especially if it enables one to escape or circumvent the power of others.[2]

The foole is a rational agent who is prepared to be unjust if it is to their advantage. Like other political philosophers, going back to Plato in *The Republic*, Hobbes sought in vain for an argument that would convince a rational agent that acting justly was always in their interest.

A hundred years later David Hume struggled with exactly the same question:

> A sensible knave . . . may think that an act of iniquity . . . will make a considerable addition to his fortune, without causing any considerable breach in the social union. . . . That honesty is the best policy, may be a good general rule, but is liable to many exceptions: and he, it may perhaps be thought, conducts himself

with most wisdom, who observes the general rule, and takes advantage of all the exceptions. I must confess that, if a man think that this reasoning much requires an answer, it would be a little difficult to find any which will to him appear satisfactory and convincing.[3]

Hume was more up-front about his inability to convince the knave that acting justly was always in his own interest.

However unsuccessful moral and political philosophy may be in convincing the knave and the foole to behave themselves, human society requires for its own protection that they be induced to do so. This is the most important job of economics.

What we need are institutions—practices, norms, rules, laws—that protect us from Hume's knave and Hobbes's foole. We need to shape and constrain their behavior in ways that attain or advance everyone else's interests—that is, the overlapping and mutual interests of the rest of us non-fooles, non-knaves.

Economics provides the best, and in fact the only, effective tools for doing this job of designing and fine-tuning the institutions we need to protect ourselves from fooles and knaves. It does so by starting out with the assumption that we are all Hobbes's fooles and Hume's knaves. That makes Milton Friedman, Gary Becker, and the economic theorists who came before and after them right to adopt rational choice theory, but for the wrong reasons. We should treat people *as if* they satisfied the RCT model of *Homo economicus*—not because most of us come anywhere close to being completely rational but to protect ourselves from those who do. That's what economists' relentless employment of the model of rationality really does for us. Humanity's need to prevent civilization being unraveled, returned to Hobbes's state of nature by the subversion of knaves and fooles, is more than enough reason for adopting the RCT model as the chief tool of institution design and improvement.

We are fortunate to already have one set of institutions that does a pretty good job of protecting us most of the time. Many real economic markets (but not the two most important ones, for labor and finance) come close enough to the model of perfect competition to do the job well enough most of the time, when the stakes are low. Perfect competition, if it ever existed, would protect us perfectly from the knave and the foole. There is nothing even they could do to subvert such a market to their advantage. Many real markets come close enough to perfect to do the job pretty well. Real markets for most low-cost items are far enough away from monopolies and monopsonies that their participants are effectively price takers and market failures do not arise. There are enough sellers within a buyer's reach, and enough substitutes, that few sellers can successfully secure large rents on most retail markets. But, as we have seen in the last few chapters, there are several key markets rife with rent seeking and price setting. These are the markets that need to be continuously policed and frequently redesigned to protect ourselves against the knave and the foole. Economic theory gives us the tools to do the job.

Thank Darwinian Cultural Selection for Spontaneous Order

The great puzzle about real markets is how an institution like the competitive price system could have emerged in the first place. This is the problem of *spontaneous order*, first fully recognized by an economist, Friedrich Hayek, in a paper that did as much to win him the Nobel Prize as any single journal article could, and without a single equation in it. In "The Social Uses of Knowledge," Hayek showed exactly what economic problem the price system solves and why that problem cannot be solved by any amount of conscious human deliberation or intentional planning, no matter how advanced our science and technology.

He didn't put it this way, but Hayek's first insight was that the market price system is an information storage, processing, and retrieval device—a virtual computer. When it is working right, the price system sends each individual just the information needed by that agent to decide what to buy and consume, what to produce and sell. It constantly updates the information we need in real time, then coordinates all our individual decisions to compute and produce an aggregate outcome that is allocatively efficient for the whole society. And it does this for all inputs and all outputs, all the time, no matter how many inputs and outputs and consumers and producers there are, without communicating misinformation or superfluous detail that no one needs to make their economic decisions.

Hayek's next great insight was to see that no individual or set of individuals could perform the calculations required to coordinate all consumers' and producers' plans. They couldn't do it, even if, per impossibile, they could send and receive the needed information to all economic participants. No amount of human intelligence can do what the virtual computer does.

In 1945, when Hayek wrote this paper, central planning had just won the Second World War for the Allies and the Soviet Union, and it was widely thought that planning could also avoid a repeat of the vast prewar depression of the 1930s that the market price system had permitted. Planning was all the rage—scientific, objective, efficient, disinterested, unlike the anarchic free market with its shortages of food and surpluses of labor. The fashionability of this idea is what made Hayek's paper so important.

He showed that the price system does what no human could attempt or accomplish: continually distribute and update to each of us just the specific information about changes anywhere in the economy that we each need to make our own economic decisions. It should be obvious that no individual ever consciously or intentionally came up with this idea of a virtual computer to solve a society's information processing problem, and still less convinced everyone to employ it.

But this raised for Hayek the problem of spontaneous order: Where did this institution that delivers exactly what humans need and can't provide for themselves come from, and what maintains its existence?

There are just four possible explanations for the emergence of the market price system: it was ordained and delivered by a benevolent deity to meet our needs; it was intentionally designed by our forefathers, who gathered together in the state of nature to hammer out a social contract; it emerged by complete accident; or it emerged through some process of Darwinian cultural selection molding our behavior to produce the solution to an environmental design problem. Of the four possibilities, there is only one we can take seriously. Hayek recognized that the price system nobody designed but that works so well for us all must be an unintended result of an undirected process of Darwinian cultural evolution. It is the product of a process of random variation and environmental selection operating on human behavior and cultural institutions over millennia, a process that started even before our evolutionary ancestors emerged onto the African savanna a million or so years ago.

The genus *Homo* was never entirely in Hobbes's state of nature. But our evolutionary ancestors must have faced something close to the anarchy in which "the life of man [is] solitary, poor, nasty, brutish and short."[4] Nevertheless, a million or so years before recorded history began, there were already practices regulating human conduct and making at least some of it cooperative and collaborative. Without these institutions the puny creatures pushed out of the forests would never have survived, let alone risen from the bottom of the food chain on the African savanna to the top. The most important of the institutions our ancestors needed to survive were the rules governing the sharing of domestic tasks and the risks and rewards of hunting and gathering.

It's a great problem in evolutionary anthropology to explain how a cooperative species like us could have emerged among all the other creatures that were the result of three billion years of relentless selection for maximizers of short-term individual fitness. Why didn't natural selection produce a whole species of Hobbes's foole and Hume's knave? Unlike all other higher mammals, including the primates, species in the genus *Homo* show marks of cooperation and collaboration in their archaeological record. Ever since Charles Darwin first pondered the matter in *The Descent of Man*, the question of how the cooperative, collaborative mode of living emerged has troubled behavioral biologists, social ecologists, and cognitive social psychologists. Most of the best answers to the question have been provided by evolutionary game theory.

Game theory in biology began by undermining Darwin's answer to the question of how natural selection produced ancestral humans who were nice to each other, prepared to reduce their fitness to enhance the fitness of others. Groups of cooperators, Darwin argued, would do better than groups of egoists who acted in their own selfish interests, because the groups of cooperators could more easily collaborate, especially in war, coordinating their actions to project force and defend against force, sometimes sacrificing themselves for one another's survival and the group's advantage. Biologists like John Maynard Smith, who introduced game theory to their discipline, recognized early on that Darwin's group selection process wouldn't work in the long run to maintain cooperative groups. Every group of cooperators would always be at risk of being subverted into a population of individuals going at each other's throats. If everyone were hardwired to play the cooperative strategy in a prisoner's dilemma, then there would be strong selection for the strategy of always defecting whenever it occurred as a random mutation. If everyone were cooperating because they were indoctrinated to do so instead of hardwired for it, then the optimal strategy would be to defect when you could get away with it, which would be enough to completely unravel cooperation over time.

Cooperation had to be an evolutionarily stable strategy, one that couldn't be subverted or invaded by noncooperators. If the problem of spontaneous order couldn't be solved at least in principle, then Hayek's explanation of the emergence of the market price system would sound like nothing more than hopeful hand waving. That's because each act of cooperation on the African savanna has the same problem as each act of market exchange. They are both prisoner's dilemmas in which the rational strategy is to not cooperate.

Let's see why.

A single trading opportunity is a prisoner's dilemma in which rationality prevents the trade from ever taking place. Each individual exchange of goods for goods, or goods for money, is a one-shot prisoner's dilemma where each party's optimal strategy is to grab what they want while withdrawing what they offer, if they can get away with it. Enforcement of exchange is spotty, costly, and ineffective, especially when parties' endowments differ prior to the opportunity for exchange. Initially, successful exchange was a fragile expectation and a high risk. Making exchange reliable was a problem that had to be solved before what Adam Smith called "the propensity to truck, barter, and exchange" could emerge.[5]

The same problem arises on the African savanna: two scavenging hominins happening upon a carcass can both fall upon it with their stone tools. But if the megafauna predator who downed the carcass returns and surprises them, they will both be mauled, if not killed. One can stand guard, taking all the risks, while the other gorges himself. Neither is willing to stand guard while the other eats. They can take turns, alternating between standing guard and feeding. But who gorges first while the other stands guard? Neither can trust the other to take turns, leaving "as much and as good" while the other keeps watch. The result is the suboptimal outcome of the prisoner's dilemma: both gorge at great risk, an outcome much worse than taking turns.

But what if the two scavengers are playing a repeated, iterated game made up of a large number of one-shot PDs? Then the rational strategy, selected for if one or both parties try it out enough times, might be something quite different from always defecting. It might enable cooperation to emerge as a matter of spontaneous order.

Economic Theory as Institution Design

Back in the 1980s, when the application of game theory to biology was just beginning, a political scientist named Robert Axelrod, influenced by the great evolutionary biologist William Hamilton, began exploring what behaviors would be selected in repeated PD games. In *The Evolution of Cooperation*, Axelrod was able to identify conditions under which competitors' best strategies result in cooperation that can't be subverted or invaded. Axelrod's laboratory was a computer, running simulations of iterated PD games over thousands of turns in which a dozen or more different strategies, designed by game theorists, competed for survival and replication.

In Axelrod's computer simulation, many strategies for playing repeated PD games were matched against each other over and over again. Some of these strategies were simple: always defect, always cooperate, flip a coin to decide, or defect every other game. Then there were some slightly more complicated strategies. In tit for tat, a player cooperates the first time they encounter another player and then does what the other player did the previous time in subsequent encounters. Another strategy, tat for tit, starts out with a player defecting and then copying the other player's last move. Then there's tit for two tats—start out nice and don't switch to defect till the other player defects twice. It's easy to multiply strategies and contrive complicated ones. Each game had a payoff matrix that obeyed the PD preference-ranking orders described in chapter 8—say, $7 to each player when both cooperate, $5 when both defect, and $10 to a defector when the other player cooperates and $1 to the cooperator. The simulation can be run any number of times. It is also possible to model natural selection for strategies that compete with one another: just add up each strategy's total winnings after ten or a hundred or a thousand rounds, then cull the strategies that have gained the least and multiply the strategies that have done the best. What Axelrod discovered in his simulations was that under a wide range of values for payoffs and a large enough number of rounds, tit for tat does better and better until the only players left are all using this strategy. Notice that when every player is doing this, every player is cooperating on every round against every other player. They might as well all be playing unconditional cooperation. But also notice that their cooperative strategy can't be invaded by a strategy of always defecting. Defection might do better the first time against tit for tat, but it will do worse in every subsequent round while tit-for-tat strategists will do better playing among each other.

Axelrod's take-home message was that when the game is an iterated prisoner's dilemma, the strategy with the best chance of succeeding won't always or even often be to defect. Under many conditions the most successful strategy will be one of conditional cooperation. His analysis of why tit for tat does better than other strategies in iterated PD was threefold: it's nice (it starts with cooperation), it's clear (other players can easily figure it out), and it's forgiving (once taken advantage of, a player using it will return to cooperating immediately if the other player does).

Tit for tat is a relatively low-risk and high-reward strategy in many iterated PD games, provided the stakes are low enough and the likelihood of repeated interaction is high enough. If the circumstances on the African savanna two million years ago were like this, there would have been strong selection for tit-for-tat strategies once they were tried, and for any other features of human psychology, such as emotions and norms, that fostered tit-for-tat strategies. Once behavior driven by such strategies was up and running, it could have provided a human environment—a niche—that would continue to select for and foster other cooperative strategies and select for the psychological predilections that make such strategies easy to employ, including the strategy of exchange.

Is this how cooperation started among our hominid ancestors? We can't know. Behaviors don't fossilize well, and there are no artifacts of collaborative hunting or crafts requiring cooperation to create until long after the initial problems of survival on the African savanna were solved. But the evolutionary game theory model Axelrod contrived is a scenario that makes spontaneous emergence of order something more than a pious hope. In principle it reconciles the institutions of human civilization with the exigencies of natural selection. It enables us to see how nature could have put in place the first and most important prerequisites for trade long before any traders had the slightest notion of how and why trade does what no individuals could do for themselves or anyone else. Long before trade became commonplace, Darwinian cultural selection had solved the PD problem each exchange presents by nesting it in another, much longer game with a cooperative solution.

It looks like trading began in earnest more than two hundred thousand years ago, by the time obsidian began to move great distances from its sources in one part of East Africa to the Acheulean workshops hundreds of kilometers away. Humanity had to already have emerged from the state of nature before that could begin to happen. Tit-for-tat strategy is plausibly part of the story of how it did so. If you don't like this story of how the cooperation that makes trade possible started in our genus, you need to provide another one compatible with selection for maximizing individual fitness.[6]

There are other instances of the emergence of Hayek's spontaneous order, institutions that confer benefits, meet needs, and secure advantages for economic agents through Darwinian cultural processes even though no one dreamed them up or consciously implemented them in order to do so.

One institution that exemplifies Hayek's spontaneous order was the subject of chapter 4: money. Money has a couple of crucial functions: solving the double coincidence of wants problem and making it possible to postpone consumption of current surpluses. To do this it needs certain properties—portability, divisibility, durability, intrinsic desirability to humans, and limited supply. We have a certain amount of cross-cultural evidence that a process of blind variation and environmental filtration searched through a "design space" that included a variety of strange candidates for the office of money till it hit upon small pieces of shiny metal in sixth-century BCE Lydia and China. This alternative quickly outcompeted other solutions to the two problems. But we know that nobody started out asking themselves, How do we solve the double coincidence of wants problem? No one recognized the problem under that name until long after shiny pieces of metal began to solve it.

Only once money was well established as a solution to a problem no one had been aware of could human intentions, recognizing and exploiting opportunities that money made possible, begin to refine the institution. Conscious design could fine-tune it, and people convinced others to join in supporting and sustaining the refined institution. The process eventually introduced *fiat currency* (pieces of paper) and all its successor vehicles for solving the problems that current forms of money deal with.

Still another example of spontaneous order is the emergence of the firm as the institution that perpetuates the division of labor. Ronald Coase, another Nobel Prize winner, suggested that the firm emerged in the seventeenth century because it solved a design problem presented by the division of labor even though no one identified the problem or consciously designed the firm as a solution. The transaction costs—mainly the risks associated with specializing in only one component of a multipart product—were high. The division of labor was unfeasible without assurance that the specialist would secure adequate reward for their share when the entire product to which the specialist contributed was completed and sold. Enter Darwinian cultural selection. Groups of individual specialists who came together for any reason in any form that increased trust and lowered risk—family, propinquity, guild, religion—would do better by way of the division of labor than others, and so would prosper, along with the strategy—the norm of organization—they discovered, no matter the reason they might have given themselves for doing so.

Spontaneous order is possible only as the result of some sort of Darwinian process operating on variations in strategic interactions between humans. The Darwinian process builds the prerequisite institutions for even the simplest economies. Understanding how and why it works is indispensable to understanding why we need economic theory despite all its foibles and weaknesses as an explanatory or predictive "science."

Incentive-Compatible Mechanisms

Evolutionary and experimental game theory can't uncover the actual trajectory that took hominid cultural evolution past the obstacle of shortsighted self-interest to cooperative institutions. But it can show how the emergence of cooperation was possible. Once humanity managed the trick, the disposition "to truck, barter, and exchange," in Smith's words, could develop. Once it did, free exchange was here to stay. That's because it comes close to solving our problem of protecting ourselves from Hobbes's foole and Hume's knave, both of whom persistently appear in human society everywhere, in every generation. The competitive market is perhaps the only permanently foole-proof and knave-proof human institution, the uniquely effective incentive-compatible mechanism.

Incentive-compatible mechanism is another Nobel Prize–winning conceptual innovation, this one courtesy of the economist Leonid Hurwicz, who had to wait till he was ninety to share the prize with a couple of younger economists who took his idea and ran with it. An economic *mechanism* is a set of rules, practices, norms, or laws that successfully produce some outcome. It's the blueprint for a procedure that harnesses other people's private knowledge and self-regarding, self-interested, knavish preferences to achieve the objective that you want. It gives people incentives that are compatible with their self-interest to act in ways that achieve the mechanism's objective, even when that objective is not shared by the people it incentivizes.

Perverse incentives illuminate incentive compatibility by contrast. The classic examples are government programs to stamp out some pest or vermin by offering a bounty on each dead varmint brought in. As Mark Twain reported in his autobiography, this invariably leads people to breed the pests for profit, thus proliferating what the government wants to minimize.

Designing an incentive-compatible mechanism requires the designers to keep in mind that people, especially Hobbes's foole and Hume's knave, will misrepresent their knowledge and beliefs if it is to their advantage to do so. They'll also want to misrepresent their preference rankings and the strength of their desires, if doing so is advantageous for them. At a minimum, feigning indifference to some good for sale may drive its price down, while misrepresenting the real cost of what one makes may give it a higher purchase price.

To work, to achieve the aims it was designed for, a mechanism has to force agents to reveal their real preferences and beliefs even when they want to hide them. It can't assume honesty. It has to treat everyone as if they were Hobbes's foole or Hume's knave.

Here's the amazing thing about the real market for most of the consumer goods bought and sold every day—the most important of the institutions that result from the Darwinian process of producing spontaneous order: In the complete absence of any designer's interest in incentive compatibility, a process of blind variation and environmental filtration produced an institution, the exchange market, that approaches incentive compatibility more closely than anything we can design. It solves the problem posed by Hobbes's foole and Hume's knave better than anything else we've come up with, even since economic theorists identified the problems of incentive compatibility in institution design and set about trying to solve them.

The perfectly competitive market makes us all price takers. It doesn't matter what we want people to think we want or what we want people to think we know; if we're price takers we have to buy and sell at the market price, which is the price that achieves an allocatively efficient, Pareto optimal general equilibrium. The competitive market is a perfectly incentive compatible institution. No one could have designed a better one for harnessing our real desires and beliefs to produce an outcome none of us aim at. No matter how we try to hide, lie about, or misrepresent them, on a perfectly competitive market our real beliefs and actual desires come out in our actions—our sales and purchases.

Wherever people have inside information, either about their own desires and preferences or about the world in which they seek to maximize these preferences, there will be incentive compatibility problems in institution design. Always! By showing us one good approximation to a solution, mother nature has laid out some baseline strategies we can use to solve other problems: make institutions as much like competitive markets as one can; privatize, but in ways that force buyers and sellers to reveal their real beliefs about their costs and benefits.

Until Hurwicz articulated the problem of incentive-compatible institution design, economists thought that the main problem an economy had to solve was one of efficiently allocating goods to satisfy preferences. Economics, in Lionel Robbins's widely accepted definition, was "the science [of] human behaviour as a relationship between ends and scarce means which have alternative uses."[7] But that missed at least half of the subject, and perhaps the more important half. Scarcity is not just a matter of insufficient means. It's also a matter of not enough correct information about means.

What society needs to achieve optimal outcomes, to protect itself against fools and knaves, are institutions that use accurate information and force everyone to reveal it when it is widely dispersed, easily hidden, and valuable to those who have it when others don't. When the mindset of economic theory was occupied by the beauty of perfect competition, the problem of incentive compatibility wasn't noticed because it was already solved. Once it became obvious how far the most important markets are from perfect competition, the problem was salient.

Economics is crucial because we are always at risk of facing strategic interaction problems, confronting knaves and fools who are prepared to lie, make misrepresentations, and pretend to comply while searching for ways to secure rents from the rest of us compliant participants in the economy. Why are we always at risk of being conned by corner-cutting knaves and untrustworthy fools? Because we all are knaves and fools to some extent, at least some of the time—some of us much more so than others, some much less so, but almost all of us at least a little bit. Like almost every human behavior, the disposition to fair and honest dealing is distributed in a nice bell-shaped curve: on the right side of the curve is the small number of us who would never cheat at Monopoly, shoplift when the checkout line is too long, underreport our taxable income, or run a stop sign in the middle of the night at an empty intersection. On the other end of the spectrum are sociopaths ready to do the most horrible things while looking us steadily in the eye, retaining our full confidence that they are just like us, almost fully trustworthy. Fortunately, there are as few sociopaths as, unfortunately, there are utterly and completely honest people. We cannot identify RCT-rational fools and knaves before they do serious damage to the rest of us. We can't even identify when and where most of the rest of us will be tempted to take advantage of the opportunities to free ride or worse. So the safest thing in designing institutions is to assume the worst and treat us all as rational agents—which is exactly what economics does. And the part of economics that is most crucial to protecting us from the worst of ourselves turns out to be game theory.

Game Theory Wins the Bidding

"Gaming the system" describes behavior we know about only too well. It's using inside knowledge about how the system works to secure some outcome you want but haven't necessarily earned. What we need are social, political, and economic institutions that can't be gamed, even by people who want to do so. How can we find such institutions—the incentive-compatible ones?

Game theory enables us to search for such institutions in a number of different ways with a number of different tools. Experimental game theorists run controlled experiments in which people's behavior is monitored as the rules of the game they play are varied. We can see how subjects change strategies and whether the outcome of the competition is optimal in some respect or another. The rules that result in the outcome the theorist wants, for example an outcome in which all players finish with equal winnings, can then be used to design an egalitarian real-world institution. Like Axelrod, some game theorists run computer simulations matching large numbers of RCT-rational bots playing against one another under various conditions a large number of times to see which strategy comes out ahead. By varying the rules they can see which strategies do best. Then they can recommend the design of institutions that enforce strategies promoting socially or privately desired outcomes when real people—including RCT-rational fools and knaves—compete. These models that can be run on a computer are called *agent-based models*. We'll see an example exploiting Axelrod's results below.

Mathematical game theory derives theorems from assumptions about rational agents, just like the economic theory that gave us the proof of Pareto optimality in the general equilibrium model of perfect competition. But game theory's proofs start with the removal of one or more of the idealized assumptions of the perfectly competitive model. Theorists often remove the assumption that everyone has complete information and knows not just the same things as everyone else but everything relevant to making a rational decision. Mathematical game theorists can derive theorems about what happens when people hide information that other players could use to improve their outcomes. This information can be about their own preferences and resources or about other people's; it can also be about what they will do once the game begins or the institution is up and running.

It's obvious how important institution design is when it's the government designing institutions to herd big players—large corporations, banks, and other financial market makers—into doing the right thing when they have every interest in gaming the system. Notice that every player can apply game theory in these contexts: the government asks game theorists to design laws to persuade firms to do something in the public interest, or (more often) to prevent them from doing things adverse to the public interest; firms use game theory to figure out how to respond, find loopholes, or take calculated risks of being caught cheating.

What mathematical game theory has revealed in proofs is not only surprising but in some cases of great value to society. In fact, one of the most valuable insights in terms of its monetary return to the government is also one of its most surprising and counterintuitive results. It's not hard to understand even without going through the math that other results require.

Mathematical game theorists have won their share of Nobel Prizes, most for showing how to design auctions. In retrospect it's not surprising that this has been the earliest arena of success, since auctions are really simple institutions in which one seller sets the rules and there are several bidders who all know the rules and know that the other bidders know them, too. It's almost the simplest case of a monopoly, one that highlights the strategic action problem facing both sides of a monopoly deal. Auctions focus the game theorist's attention on the crucial information asymmetry. The bidders know how valuable whatever they're bidding on is to them, but they don't know its value to other bidders. Because the number of bidders can be small, it's easier for bidders to collude, whether by explicit agreement or via "a wink and a nod" implicit collusion. This will be especially tempting and easy if they know several items will be auctioned off sequentially. The economic problem for the owner of the goods to be auctioned is how to design the bidding rules to get the highest price for each item.

When the US government began to sell parts of the electromagnetic spectrum to mobile-telephone companies all over the country, it faced the problem of designing the auction to maximize the return for the owners of this commodity: taxpayers. It needed a set of rules that would force phone companies to bid up the prices to the full values of the slices of the spectrum they would own. It needed to prevent companies from forming explicit or implicit coalitions to bid in collusive ways to divide up the spectrum at a lower cost than they would pay if they each acted independently. The government turned to game theorists to solve its auction design problem.

Recall from the discussion of monopoly models that the monopolist's problem is determining the slope and intercept of the demand curve, which the price-taking seller doesn't have to do in perfect competition. Meanwhile, the buyer has to decide whether to buy at the price the monopolist seller sets or refrain in the hope the seller will lower the price. Because an auction is a monopolist's market, both of these problems arise. But the buyer faces the additional problem of outbidding other buyers. How much to bid? Not overpaying is easy: just stop bidding once a bid exceeds your private valuation. Every bidder knows that all the other bidders will bid less than their private valuation. That means the seller faces the prospect of leaving money on the table: the winning bid will always be less than, or at most exactly equal to, the winner's full valuation of the item up for auction (if the bidder can figure out exactly what that amount is). The auctioneer needs to contrive a bidding rule that forces bidders to bid right up to their true valuation. One game theorist, William Vickrey, began deriving theorems proving that some auction rules solve the seller's problem of forcing bidders to reveal their real valuations in their bids. Vickrey's surprising, counterintuitive mathematical result is that the problem is solved in an auction in which the highest bidder has to pay only the amount the second-highest bidder offered plus one cent.

What Vickrey proved was that the *dominant strategy*—the best thing one can do, no matter what others choose—in a *second-price auction* with a single, indivisible item to be sold is to bid your full value of the item. We can run through the reasoning to get this counterintuitive result.

If I bid my full value but someone else values the good more and bids their true value, they win, and welcome to it! I would have had buyer's regret if I had outbid them and offered more than the item is worth to me. If I bid less than my true value and another bidder wins with a bid higher than mine but less than my true valuation, I'll lose the bid and be worse off than I could have been. If I bid my true value and it's greater than anyone else's true value, then in a highest-bid auction I will have bid more than I needed to. In a top-price auction all bidders have to worry that they will overbid the competition and so they each have an incentive to bid below their true value.

In a second-price auction, even if I radically overbid my value, if I win I won't pay the amount I bid but the amount the second-highest bidder offered (plus a penny). Every bidder knows this. What if all bid recklessly, knowing they won't have to pay their bid but only the second-highest bid? Well, the second-highest bid was reckless, too, and the winner will likely be burned even paying that. If everyone knows this and knows that every other bidder knows it, too, each will bid exactly what they think the item is worth to them—no more and no less. They will reveal their true valuations in their actual bids. And that's what the seller wants!

The design of auctions by game theorists has enabled governments all over the world to sell scarce mobile-telephone and broadband spectra in ways that discourage collusion and secure higher returns to owners. The second-price auction also became the staple institution for auctioning advertising space on the internet soon after the market became competitive.

Of course, once an auctioneer sets up the rules for an auction, bidders have an incentive to figure out if there is any way, by acting alone or with others, to win the auction at some amount less even than the second-place bidder's true value. This arms race between parties to a series of strategic interactions is characteristic of every real economy's multiple arenas for rent seeking.

There are many complicated problems of mechanism design. Auction design is among the easiest and so has seen the most mathematically developed results. But there is a more complicated design problem faced by every small business every time an employee is hired. It grows more serious as businesses become larger and begin to require professional management. In every firm, owners face the *principal-agent problem*: how to align the employees'—the agents'—preferences with those of the owners—the principals. The owners of a retail store seek maximum profit. The salesperson hired seeks minimal work, late openings, early closings, and opportunities to benefit friends and family in purchase opportunities. If the salesperson is a real knave or foole, they may also engage in a little larceny from the till. How to structure the salesclerk's incentives to match the owner's interests? It's an institution design problem. But it's much more serious when millions are at stake in a massive capital project. This is one area where the mechanism design theorist's expertise can be put to work.

One Cheer for Capitalist Inequality, One Cheer for Socialist Equality?

Here's an example cribbed from the 2007 Nobel lecture of Roger Meyerson, one of the economists who won the prize with Hurwicz for developing mechanism design theory.[8] It illuminates several dimensions of the principal-agent problem and how it might be solved.

In the production of expensive capital goods that require a lot of inputs in the form of labor and equipment and their organization over long periods of time, managers have substantial opportunities to misuse, divert, and steal inputs; secure bribes from suppliers; suborn other employees; and engage in other nefarious activities. For these and other reasons the success of the production process hinges on the manager's performance. If the difference in outcomes between good and bad management is large enough, the outcome of the project will hinge on the manager's choice of good versus bad strategies of personal performance. (*Good* here means good for the owners.) The problem for those who provide the capital to undertake and complete the project is to ensure the manager chooses to be good at the job. How to do this? Pay the manager enough to make it worthwhile for them to administer well. How much will this be? This wage will have to be a rent, an amount greater than the marginal productivity of the manager's labor. Why? Because in this market perfect competition doesn't prevail: there is price setting, a strategic interaction problem. The manager's wage, *w*, must satisfy a *moral hazard constraint*. That is, it has to be large enough to outweigh the benefit to the manager of managing badly. If a manager puts up some amount of their own capital for projects they run, then the expected wage payout for good management minus what the manager puts up will have to exceed the benefit to the manager of mismanagement—bribes, kickbacks, diversion of inputs—along with the expected wage payout for mismanagement, minus what the manager will lose in the capital they put up for the project.

We can write this down as an equation using elementary algebra. But first let's use words instead of symbols to make the equation easier to keep in mind.

Start with the probable net benefit to the manager of good management:

[The probability of success with good management multiplied by the manager's wage] − [the probability of failure of good management multiplied by the manager's investment in the project].

This amount must be larger than what the manager would probably attain with bad management:

[The payoff to the manager from bad management] + [the probability of success with bad management multiplied by the manager's wage] − [the probability of failure with bad management multiplied by the amount the manager invested].

In symbols:

$p_G w - (1 - p_G)A \geq B + p_B w - (1 - p_B)A.$

The p's are probabilities of success, with subscripts G for good management and B for bad management. The manager's wage is w, A is what the manager invests in the project from their own resources, and B is what the manger can gain from bad management.

To begin with the manager needs to be paid a wage that will induce them to take the job. This is the *participation constraint*. The wage times the probability of the project's success under good management has to be greater than the chance that even with good management the manager's stake will be lost.

In symbols:

$P_G w - (1 - p_G)A \geq 0.$

Of course the manager's wage has to be at least as high as what the manager invests in the project from their own assets, since once invested they will be irretrievable. That is represented as

$w \geq -A.$

It's $-A$ since A is what the manager invests and therefore $-A$ is what the manager stands to lose if the project fails.

Subject to these constraints, the payoff to the owners (or the government, for public works projects) of successful completion can be calculated. In words again, it will be the probability of success with good management multiplied by the net return to the owners, minus the revenue and minus the manager's wage, plus the probability of failure even with good management multiplied by the manager's capital investment, minus the total capital investment needed to complete the project. In elementary algebra, that is

$$V = p_G(R - w) + (1 - p_G)A - K$$

where V is the net profit to the owners, government, or society as a whole, R is the revenue returned by the completed project when put in service, and K is the total capital amount needed to accomplish the project.

If we plug in some numbers, it's easy to figure out how much to pay managers and how much to require managers to contribute to projects from their own pockets. That amount is what will give the manager a sufficiently strong incentive to manage well and will make the net benefit of the finished project a positive amount.

There is some bad news for egalitarianism in the search for incentive-compatible mechanisms that successfully produce capital projects while minimizing graft and corruption. If A, the amount the manager has available to contribute to the project's capital, isn't very large, then the expected value of the payoff to him of bad management (taking bribes, skimming off the top, etc.) may be much greater:

$$A < Bp_G/p_G - p_B.$$

That means the manager's wage must be (much) larger than the potential payoff to corruption and mismanagement minus his (small) investment. The owners therefore will have to give up a significant amount of revenue to pay a high enough wage to ensure good management.

The moral of this game-theoretical story is particularly sobering for socialism or any set of institutions that ensures equality of income and wealth. In such societies, no one has access to a large value for A, the resources one might contribute to a capital-intensive project. Therefore, to ensure completion of such projects, society must abide high salaries for managers, undermining its egalitarian commitments. It's the only way to minimize the manger's *moral hazard*—the temptation to skim off the top and slack off on the job. Game theorists can argue that given reasonable values for the variables in their models there's no way to deliver economic growth without persistent inequalities in income that will snowball into inherited (and therefore ever-increasing) inequalities of wealth.

But there is worse to come in thinking through the implications of the design problem society faces. It helps us understand Joseph Stalin's abuses in his five-year plans to industrialize the Soviet Union and makes intelligible aspects of the great terror he visited on the Soviet Union's managerial elite.

Incentive compatibility can be a matter of sticks as well as carrots. Add to our variables z, representing different or graduated punishments for failure by managers. Punishment won't be as valuable to the rest of society as seizing managers' assets in case of failure, but in an egalitarian society without much wealth, there are no assets worth seizing. Those with authority over managers need to find a combination of wages and punishments that maximizes the prospects of successfully completing the capital project. The wage, w, will have to be greater than A, what (if anything) the manager contributes, and the punishment will have to be nonzero: $z > 0$. There will be a *moral hazard constraint* expressed in an equation like the moral hazard constraint equation for capitalism, except that for A we will have to substitute $(A+z)$—the sum of the manager's investment and the punishment for failure. What about the *participation constraint*? If we assume that Stalin could force people to do jobs, he wouldn't have needed to incentivize people to take on managerial jobs to satisfy a participation constraint. Stalin could maximize net profit to the economy, V, simply by setting the manager's wage at zero and choosing the right punishment value for z. Or he could adopt a mechanism that both created a managerial elite by offering better wages to managers and subjected them to harsher penalties for failure to achieve the plan. That's what he did.

The take-home message for solving the principal-agent problem, at least for CEOs, is to minimize moral hazard by designing the institution of CEO jobs to allow for high pay but demand metaphorical skin in the game. Exactly how much of each to incorporate in the rules is a matter for experimentation, since potential CEOs will themselves have to make some calculations and experiment in their own performance to find their optimal strategies.

What about the adverse selection problem of a wage so high that it attracts bad managers prepared to lie about their skills? Potential CEOs know whether they are good managers or not. But this knowledge is private. Others don't know, and all managers will assure employers that they are good. Game theory shows that this problem is intractable: you can't design an institution that will force bad managers to reveal their actual motivations. The math employs all the same variables as the moral hazard case we just walked through.

In a perfectly competitive economy, where everyone is a price taker, there are no rents and no expected profits to anyone, so V, the expected economically unjustified profit to investors in the project, will be zero. It turns out that if A, the amount the manager has to put up, is also zero, the math shows that there is no way to ensure high-quality management. But the same math also shows that a centrally planned economy, such as an egalitarian socialist one, can solve this adverse selection problem. In this economy no one has any excess capital to invest, so $A = 0$. The society pays the whole cost of a capital project. To ensure that no potential manager lies about their management qualities and that only good managers apply to be the CEO of a project, the government sets the wage at the zero rent level. It pays managers just enough to compensate them for their management effort with no rent premium. Under this condition, bad managers won't apply to do the work as there is nothing in it for them.

Egalitarian socialism turns out to have serious moral hazard problems—skimming off the top—and capitalism has serious adverse selection problems—hiring crooks to begin with. Don't expect game theory to solve the principal-agent problem once and for all. The problem is a continual arms race as principals adopt new strategies and agents respond with new tactics of their own. Game theory is the tool both parties use to plot their approaches.

Designing International Trade Policies

Axelrod's project was to use evolutionary game theory to show how cooperation could have emerged among our Pleistocene ancestors. His model also works, and maybe does even more, if we fast-forward to the way international trade is organized by participants in the World Trade Organization (WTO).

Three Dutch game theorists, Sebastian Krapohl, Václav Ocelík, and Dawid Walentek (KOW for short), adopted Axelrod's approach to modeling countries' tariff policies in ways that could help governments respond to the continuing arms race that is international trade.[9] There are more than a hundred active trading nations in the WTO. It has been widely supposed for two hundred years or more that, other things being equal, more often than not free trade between pairs of countries is beneficial to both—more beneficial than erecting tariff barriers to each other, though less beneficial for one country than being able to exploit another country's openness to trade while imposing tariff barriers itself. A classic prisoner's dilemma!

The proviso of other things being equal reflects the fact that the payoffs to each strategy depend on differences between countries in the size and character of their internal markets—large or small, developed or developing, predominantly manufacturing or agricultural. But international trade is an iterated game between individual countries. That's how the WTO is organized. Cooperation or defection, bilateral free trade or trade war: states can choose, switch, and adopt different strategies with different countries. Using WTO data, KOW computed a payoff matrix for each of 130 countries (the other members of the WTO didn't report enough data to construct their payoff matrices). To model an iterated PD between two countries, they plugged the data into equations with four variables for each country: export industry as a share of its gross domestic product (total exports divided by GDP), distance between the countries, rate of protection against imports (trade-weighted tariffs on imports), and market size (GDP). Recall from chapter 8 that the payoffs in a PD don't have to be mirror images so long as the PD inequality holds:

Country A's preference ranking: I > II > IV > III
Country B's preference ranking: III > II > IV > I

Now KOW were ready to run an Axelrod-type simulation of natural selection among four strategies that countries could adopt in their bilateral trade negotiations: unconditional cooperation (unilateral free trade with every other country), unconditional defection (tariff barriers against all other countries), tit for tat, and a strategy they called generous tit for tat, in which a player retaliates against defection 70 percent of the time. In calculating payoffs, KOW deducted 5 percent from each country's payoff when playing a strategy that requires monitoring other countries' policies. In a world where everyone else is playing tit for tat, the optimal strategy is unconditional cooperation—unilateral free trade—since a country playing this strategy doesn't have to monitor other countries' behaviors. In the simulation, countries started out choosing strategies randomly and, as in Axelrod's

Economic Theory as Institution Design

original simulation, switched to strategies proven fittest on the previous round. KOW added a small mutation factor that switched countries to different strategies randomly. The fitness of a country was determined by its total payoffs against all other countries in the previous round of negotiation and trade.

After running their model through many rounds of fifty thousand turns, over and over again, starting with a variety of random distributions of strategy pairs among the 130 countries, KOW noticed some interesting patterns. First of all, in none of their runs did a strategy emerge as a clear long-term winner. The frequency of each strategy's payoff and usage waxed and waned depending on the distribution of other strategies in play. In a world

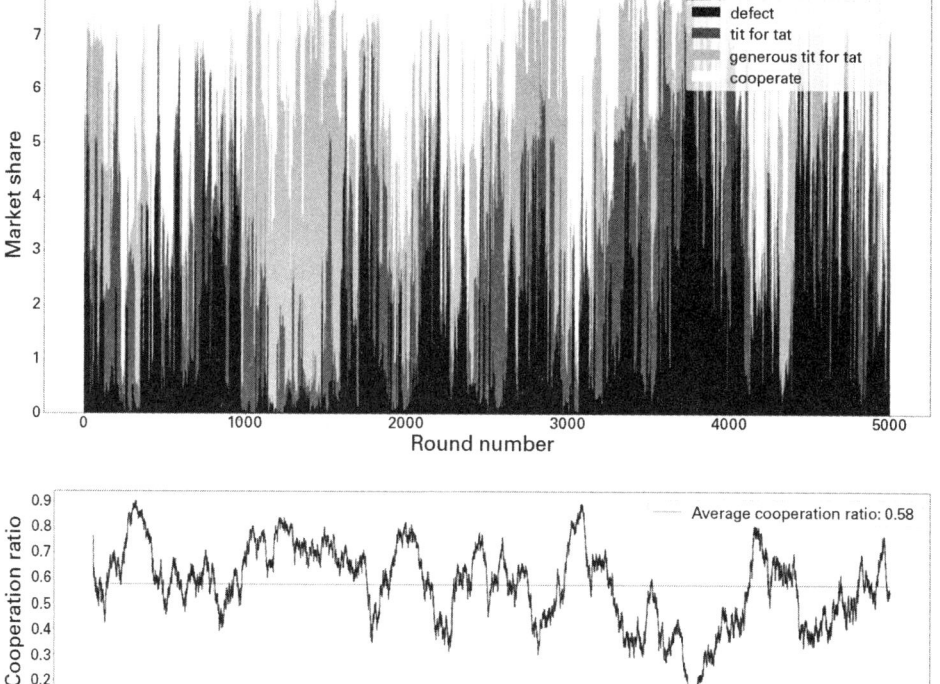

Figure 10.1
The upper graph shows the number of times per round that each of the four strategies is employed. The lower graph charts the frequency of operation among countries.

of tit-for-tat trading, unconditional cooperation becomes a more frequent strategy since it does better not having to pay the costs of surveillance. But once tit for tat becomes widespread, defection becomes the optimal strategy. When it does, the door is opened for pairs of nations that play tit for tat and we end up back where we started. Here is graph of one of KOW's fifty-thousand-turn simulation runs (each of their simulations required almost 420 million computations).

As the graph illustrates, every time a strategy predominates, other strategies increase in frequency since they can take advantage of its predominance, just as one would expect from rational players. Levels of cooperation are largely determined by which strategy is predominant at a given point in the simulation. The persistence of this cycle over a large number of simulations that select for optimal strategies among 130 real countries with very different sizes, economies, and geographic locations is a robust result.

There are obvious lessons in this evolutionary game-theoretical model for institution design and designers in world trade and the WTO. KOW's model shows us that, at least sometimes and perhaps most of the time, trading strategies might alternate in an arms race forever. It's not just that the progression of moves and countermoves between players results in new strategies; often old ones just recycle. There may not be a permanently foole- and knave-proof mechanism. What this means in institution design is that we can't let the perfect be the enemy of the good. We can't aim to permanently eliminate some option the foole and knave seek to exploit. It will have to be enough to stop them in their tracks for a while. KOW tell us that trading nations' best response to economic nationalism is to have credible retaliatory threats and use them. Their model will actually identify the countries to watch out for and quantify the estimated payoffs each country secures by trading among the 130 countries in the WTO for which there are data. There is the promise of something more than stylized facts in this approach to economic behavior.

Even Nice Guys Need Incentive-Compatible Institutions

The proportion of knaves and fooles among physicians is probably lower than in the general population. Let's hope so, anyway. But in Britain and the United States throughout the first half of the twentieth century there was a classic incentive-compatibility problem that almost all of the

Economic Theory as Institution Design

most-qualified medical students faced at the very beginning of their careers, and another incentive-compatibility problem for hospital administrators created by it. By some reasoning and experimentation, and without the aid of game theory (which hadn't been invented yet), the students and administrators solved their problems. Later, mathematical game theorists made a magnificent contribution to expanding the solution till it became a lifesaving tool in health care worldwide. Their mathematical proofs provided great benefit to many other people facing life-or-death versions of the doctors' conundrum.

The original problem was matching graduating medical students with hospitals where they would do their internships and residencies. As early as the 1940s all graduating MDs wanted to go to the best, most prestigious hospitals in their respective specialties. All the hospitals (or their administrators) wanted the best graduates. The med students' problem was that their first offers were rarely from their first choices, but if they didn't accept the first offer to arrive they might have to settle for an even less attractive choice if their preferred spot didn't appear. Hospitals had a related problem. If they gave their first choice much time to decide, they might lose that student to another hospital and then not get their second choice either because they waited too long.

The solution, adopted in the early 1950s, was the *match system*. Graduating med students listed their residency preferences in order. Hospital administrators made ordered lists of their preferences among med students. Where there was a first-choice match between a student and a hospital, the match was made and set aside for the moment—not finalized or even announced. With some students and some hospital vacancies set aside, there could be new matches. These were again set aside and the process started again. At some point an administrator would come across the name of a med student still unmatched whom they liked better than one they originally paired with, but who didn't rate them as high as the student they already claimed. At this point the hospital would switch its match to the newly discovered student. This med student couldn't do better than the hospital that switched for him or her. The student put back into the hopper also couldn't do better than be matched with one of the hospitals using their rankings of still available students. And so the process went until all acceptable med students were matched to their most preferred hospital among those that preferred them the most among available candidates. Then, and only then, once this process was finished, was the rabbit pulled out of the hat: every student got the most preferred assignment among the hospitals that expressed a preference ranking for the student; every hospital got its most preferred available student out of the ones that ranked them. No student or hospital had to grab a less desirable alternative out of worry that the more desirable one they were waiting for wouldn't turn up at all.

This system has continued in operation since 1953 and has spread, especially to other health care professions and providers. By the 1960s and '70s game theorists had taken note of this sweet solution to med students' and hospitals' interconnected problems. Without knowing about the problem and solution, game theorists had already been working on its general form, calling it the *top trading cycle*. It got that name because it makes the obvious trades first, then removes them from the market and allows a new top trade if there is one in the next cycle, and so on till all the trades people want to make are made.

Game theorists had already set to work proving mathematically that implementing a top trading cycle was an incentive-compatible institution.[10] First two game theorists, Lloyd Shapley and Herbert Scarf, proved that in the match system there is no subset of med students or hospitals that can do better than they do when everyone participates in the match. Dropping out or setting up a smaller, elite match system won't improve anyone's outcome. Then Alvin Roth showed that if everyone's preferences are *strict*—there are no ties or indifference rankings between two candidates or two hospitals—the outcome is uniquely best, a Pareto optimum in which no one can be made better off without making someone else worse off. Most importantly, he showed that each participant giving their true preferences is the dominant strategy. Recall that in the prisoner's dilemma everyone using the dominant strategy results in a suboptimal outcome for everyone. In the match system, everyone's dominant strategy produces the best possible outcome. The match system, and any top trading cycle, is a completely incentive-compatible algorithm. It can be implemented by a computer and not even a fool or a knave can successfully game it.

Mathematical game theory's formal results—Shapley and Scarf's top trading theorem and Roth's proof that it was incentive compatible—gave the designers of other institutions confidence to solve even bigger and more pressing problems than hospital residency matching. In the 1980s top trading cycles were adopted to deal with the problem of school choice in New

York and Boston schools. With many different schools to apply to, parents wanted to maximize their kid's chances of attending a good school, but knew the best schools would have the largest number of strong applicants. Many canny parents made their kids apply to their second or third choice because it was less competitive, giving up a chance at the best school for the guarantee of getting into an acceptable one. Schools had the opposite problem: they might give a spot to a satisfactory student when somewhere on their lists was a better student who wanted to enroll. With mathematical proof that a top trading cycle would produce a foole-proof (and knave-proof) match of every student with the best school that wanted them, it wasn't hard to get school administrators to adopt a top trading cycle algorithm.

At this point Roth, who had proved one of the incentive-compatibility theorems for top trading cycles (and slightly different *differed acceptance algorithms*) and had convinced school boards to adopt them, realized there was another problem that game theory could address. It was so salient that his work on it earned him the Nobel Prize, this time an underpayment for the contribution an economic theorist has made to our society.

There are many medical issues raised by kidney transplantation. But the obvious one is scarcity: many more patients need kidney transplants than there are postmortem organ donors. There is a waiting list of patients who need kidneys. But some on the list have willing potential donors—relatives and friends. Many friends and family members, in fact most of them, are ruled out as donors due to immunological and blood type incompatibilities. But what if willing donor and recipient pairs could trade kidneys that didn't work for them with others facing the same problem? Allowing such exchanges would shorten the waiting lists. It wouldn't make anyone else on the list worse off by lengthening their wait time for an organ. In fact it would make everyone still on the list better off by leaving fewer recipients between them and the top of the list. Working with physicians, health care administrators, and other economists, and employing variants on the top trading cycle algorithm, Roth designed an incentive-compatible institution to make such trading possible. The result has been thousands of lives extended, saved, and improved.

The bare bones of the mechanism Roth designed are these. Everyone on the waiting list identifies their most preferred kidney donor—the best medical match. Find all the top trading cycles of pairs of willing donors and kidney needers this preference ranking identifies. Operating on two donor-needer pairs requires four nearby operating rooms—two to remove kidneys from the donors and two to transplant kidneys into the recipients. Three pairs need six operating rooms. The next step is to make all the swaps possible given the number of contiguous operating rooms. Once all the top trading cycles have been completed, some remaining patients' preferences will have to change, as their most preferred alternative is no longer available. Shapley and Scarf's theorem establishes that no one on the waiting list can do better by organizing a coalition of donors and needers to trade among themselves. That gives all donors and needers rational confidence to join the waiting list and participate in the exchange.

Now add to the mix some donors not paired with needers—cadaver kidneys, or unconditional altruists ready to contribute a kidney without any control over who gets it. Such a donation creates one or more chains of preferences instead of a cycle: it's a linked list connecting each needer to one of the kidneys still available, including the new, unpaired one. The chains may be long or short, and all include the unpaired kidney. Giving that kidney to a needer who prefers it to others on the chain frees the needer's donor partner to give a kidney to another needer on the chain. If this produces a cycle on the chain, perform it. Take the needer at the top of the waiting list and study the chain of preferences that begins with that patient. (You could also start with the longest chain in the waiting list instead of the one that starts with the highest-priority needer). If there are cycles in that chain, make the trades. Remember no one can be made worse off by such trades. Then see how the chain has changed. Are there new cycles? Remove them. Repeat. Roth proved this chain can't be gamed by any participant, even if one were to exclude the highest-priority patient and go to the second highest, and so on. Notice that when the system employs chains in addition to cycles, an altruistic donor can expect to have a cascading benefit for many needers, not just the one receiving the altruist's kidney. The Roth scheme is not just incentive compatible but incentive creating, encouraging potential altruists to volunteer.

The presence of such altruists reduces a problem posed by Hobbes's foole and Hume's knave. Cycles of simultaneous surgeries are limited by the number of contiguous operating theaters. When the surgeries are simultaneous, there's almost no room for anyone to renege. But if they aren't simultaneous, knavish donors can wriggle out of the deal, refusing to give a kidney to the next needer in the chain once their needer got a kidney. No

Economic Theory as Institution Design

one has figured out a way to avoid this threat when enforceable contracts for organ donation and exchange are ruled out as illegal. (They are almost everywhere.) Top trading cycles and chains can't solve this problem, but they can prevent its worst-case scenario. When some knave or foole breaks the chain, at least the needer has a partner with a kidney and the chance to find themselves on a new chain with new cycles, perhaps even a cycle that includes them.

Roth's top trading cycle and chain algorithms are not easy to explain, and the proof of their incentive compatibility and optimality don't hold the attention of every transplant physician, hospital administrator, health insurance executive, or health care policymaker. One could wish they did, and that the profit-driven participants in health care could be convinced or constrained to adopt the algorithm, even at the cost of some of their revenue. But meanwhile, game theory is saving, lengthening, and improving lives in ways that help attain Adam Smith's benevolent ends, if not by the means he thought we needed to use.

Notes

Chapter 1

1. John Maynard Keynes, *The General Theory of Employment, Interest, and Money* (London: Macmillan, 1936), 383–384.

2. Thomas Hobbes, *Leviathan*, ed. E. Curley (Indianapolis: Hackett, 1994), chap. XIII, para. 9, p. 76; chap. XIII, p. 78, fn. 9.

3. Thomas Hobbes, *Leviathan* (London, 1651; Project Gutenberg, 2002), chap. XV, para. 4, https://www.gutenberg.org/files/3207/3207-h/3207-h.htm; David Hume, *Enquiry Concerning Principles of Morals* (London: A. Millar, 1751), sec. 9, pt. 2.

Chapter 2

1. Not to be confused with randomized control trials, also often abbreviated as RCT.

2. Adam Smith, *The Wealth of Nations* (London: Strahan and Cadell, 1776), bk. 1, chap. 2.

3. Smith, *Wealth of Nations*, bk. 4, chap. 2; emphasis added.

4. Milton Friedman, "The Methodology of Positive Economics," in *Essays in Positive Economics* (Chicago: University of Chicago Press, 1953), 8.

5. Gary S. Becker, "Irrational Behavior and Economic Theory," *Journal of Political Economy* 70, no. 1 (1962): 2; emphasis added.

6. The theory of the firm is just the application of rational choice theory from the description of consumer demand to the producer/seller of what the consumers can purchase. It's equally unrealistic and idealized.

Chapter 3

1. Adam Smith, *The Wealth of Nations* (London: Strahan and Cadell, 1776), bk. 1, chap. 2.

2. Adam Smith, *The Theory of Moral Sentiments* (London: Andrew Millar, 1759), 10.

3. Note that behaviorism in economics is not behavioral economics. This emerging subdiscipline requires the repudiation of twentieth-century behaviorism in psychology. It takes the mind seriously as relevant to economics. Behavioral economics rejects behaviorism and seeks the mental causes of choice. That's confusing.

4. At this point one can show how actual choices combined with revealed preferences will enable the economist to construct the rational agent's expectations—their beliefs about the probability of alternative outcomes. See chapter 5.

5. The label was first given to the theorem in Kenneth J. Arrow, "Uncertainty and the Welfare Economics of Medical Care," *American Economic Review* 53 (1963): 942.

6. Smith, *Theory of Moral Sentiments*, 350; emphasis added.

7. Lionel Robbins, *An Essay on the Nature and Significance of Economic Science* (London: Macmillan, 1932), 16.

8. Smith, *Wealth of Nations*, bk. 1, chap. 2.

Chapter 4

1. The Cobb-Douglas function was named for two economists who guessed at this form for utility in the 1930s, although marginalist utility economists had thought of it in the nineteenth century. Douglas later became a famous US senator from Illinois.

Chapter 5

1. John Maynard Keynes, *A Tract on Monetary Reform* (London: Macmillan, 1924), 80; emphasis in original.

2. John Maynard Keynes, "The General Theory of Employment," *Quarterly Journal of Economics* 51, no. 2 (1937): 213–214.

3. John Maynard Keynes, *The General Theory of Employment, Interest, and Money* (London: Macmillan, 1936), 152; emphasis in original.

4. Keynes, *General Theory of Employment*, 152.

5. Keynes, *General Theory of Employment*, 161–162; emphasis added.

6. Robert E. Lucas and Thomas J. Sargent, "After Keynesian Macroeconomics," *Federal Reserve Bank of Minneapolis Quarterly Review* 321 (Spring 1979): 1.

Chapter 6

1. Robert E. Lucas and Thomas J. Sargent, "After Keynesian Macroeconomics," *Federal Reserve Bank of Minneapolis Quarterly Review* 321 (Spring 1979): 1.

Notes

2. Robert J. Barro and Vittorio Grilli, *European Macroeconomics* (London: Macmillan, 1994), 18.

3. John Maynard Keynes, *A Tract on Monetary Reform* (London: Macmillan, 1924), 80.

4. Lucas and Sargent, "After Keynesian Macroeconomics," 8–9; emphasis in original.

5. Lucas and Sargent, "After Keynesian Macroeconomics," 8.

6. Barro and Grilli, *European Macroeconomics*, 8.

7. N. Gregory Mankiw, *Macroeconomics* (New York: Worth, 2010), 418.

8. Cynthia Misak, *Frank Ramsey: A Sheer Excess of Powers* (New York: Oxford University Press, 2020).

9. Some later models increase the types of households, for example dividing them into "young" ones that save and "old" ones that dissave, but still impose intertemporal utility maximization and market-clearing requirements.

10. David Romer, *Advanced Macroeconomics*, 2nd ed. (New York: McGraw-Hill, 2001), 60.

11. Keynes, *Tract on Monetary Reform*, 80.

12. Barro and Grilli, *European Macroeconomics*, 15–16; emphasis added.

13. J. Duesenberry, E. Fromm, and L. Klein, *The Brookings Quarterly Econometric Model of the United States* (Chicago: Rand McNally, 1965).

14. Lucas and Sargent, "After Keynesian Macroeconomics," 1.

15. Lawrence J. Christiano, Martin S. Eichenbaum, and Mathias Trabandt, "On DSGE Models," *Journal of Economic Perspectives* 32, no. 3 (2018): 114.

16. Robert Solow, "The State of Macroeconomics," *Journal of Economic Perspectives* 22, no. 1 (2008): 243.

17. Robert Lucas, "Macroeconomic Priorities," *American Economic Review* 93, no. 1 (2003): 1.

18. Christiano, Eichenbaum, and Trabandt, "On DSGE Models," 128.

19. Joseph R. Stiglitz, "Where Modern Macroeconomics Went Wrong," *Oxford Review of Economic Policy* 34, no. 1–2 (2018): 70–71.

20. Stiglitz, "Where Modern Macroeconomics," 74.

21. Stiglitz, "Where Modern Macroeconomics," 90; emphasis added.

Chapter 7

1. Adam Smith, *The Wealth of Nations* (London: Strahan and Cadell, 1776), 152.

2. Alessandro Bonnan and Stephan Lopez, "Walmart and Local Economic Development," *Economic Development Quarterly* 26, no. 4 (2019): 285–297.

Chapter 8

1. Adam Smith, *The Wealth of Nations* (London: Strahan and Cadell, 1776), 152.

2. Thomas Hobbes, *Leviathan*, ed. E. Curley (Indianapolis: Hackett, 1994), chap. XIII, p. 78, fn. 9.

3. David Sloan Wilson, Elinor Ostrom, and Michael E. Cox, "Generalizing the Core Design Principles for the Efficacy of Groups," *Journal of Economic Behavior & Organization* 90 (2013): S22.

4. Some time before Ostrom did her work, one of the founders of mathematical game theory, Reinhard Selten, proved a neat theorem: given the usual assumptions economists make about firms seeking maximum profits, coalitions of four or fewer identical firms are very probably permanently stable, coalitions of six or more are permanently unstable, and five is a point of instability. See Reinhard Selten, "A Simple Model of Imperfect Competition, where 4 Are Few and 6 Are Many," *International Journal of Game Theory* 2 (1973): 141–201.

Chapter 9

1. About the only thing us retirement savers and small investors can do to protect ourselves from being victimized by financial market rent seeking is to spread our bets—our investments—across the widest range of large capital goods producers and market-making financial institutions as we can. Buying stock market index funds allows the small seller of postponable consumption to own a little bit of all the price-setting rent seekers, but not as much as their investments could provide. Index funds with low fees, limited salaries for managers (i.e., transaction costs), and caution about risks cater to small investors' loss aversion preferences. In doing so they forgo some rents and provide middlemen with opportunities to scoop these sums up by trading with the index funds. But it's better than being taken by the financial market insiders.

2. Matt. 26:11 (New International Version).

3. There are some historical exceptions to this generalization worth mentioning because they are rare and not to be wished for even by the most committed egalitarian. Walter Scheidel's *The Great Leveler* traces several thousand years of history in which war, revolution, government instability, and disease have so wrecked civilizations as to topple even the strongest redoubts of the wealthy elites. And Thomas Piketty's *Capital in the Twentieth Century* explains the reduction in inequality during the middle half of the twentieth century as the two world wars destroyed the capital

stock of the developed world and so the source of the wealthiest people's income. But the rest of Piketty's important book proves the rule that the rich will always be with us and will always get richer.

Chapter 10

1. Thomas Hobbes, *Leviathan* (London, 1651; Project Gutenberg, 2002), chap. XV, para. 4, https://www.gutenberg.org/files/3207/3207-h/3207-h.htm.

2. Here is a modernized interpretation of Hobbes's passage, provided by an important historian of philosophy, Jonathan Bennett, that makes its point even clearer to our eyes:

> The foole has said in his heart, There is no such thing as justice, sometimes even saying it aloud. He has seriously maintained that since every man is in charge of his own survival and welfare, there could be no reason for any man not to do anything that he thought would conduce to that end; so that making or not making covenants, keeping them or breaking them, is not against reason if it conduces to one's benefit. He isn't denying that there are covenants, that they are sometimes broken and sometimes kept, and that breaches of them may be called "injustice" and the observance of them "justice." But he is suggesting that injustice may sometimes have on its side the reason that dictates to every man his own good, especially when the injustice conduces to a benefit that will enable the man to disregard not only men's dispraise and curses but also their power. (From Jonathan Bennett, *Leviathan*, 15.4, accessed January 17, 2024, https://www.earlymoderntexts.com/assets/pdfs/hobbes1651part1_2.pdf.)

3. David Hume, *An Enquiry Concerning the Principals of Morals* (London: A. Millar, 1751), sec. 9, 22–25.

4. Thomas Hobbes, *Leviathan*, ed. E. Curley (Indianapolis: Hackett, 1994), chap. XIII, para. 9, p. 76.

5. Adam Smith, *The Wealth of Nations* (London: Strahan and Cadell, 1776), bk. 1, chap. 2.

6. In *Second Nature: Economic Origins of Human Evolution*, Haim Ofek offers a powerful and learned argument that the emergence of trade drove much of human evolution from the earliest emergence of humans on the African savanna. Ofek has a particularly striking argument that the kind of commodity that was first traded was the *good idea*, a commodity that is consumed non-rivalrously but excludably. In particular Ofek suggests that making and keeping fires going was one such idea. We saw in chapter 8 why competitive markets fail to produce good ideas at anything like optimal levels because they are non-rivalrous in consumption and hard to guard. The warmth a fire provides in a cave on a cold night, however, is something from which it may be easy to exclude others, so making it private property by keeping the technology secret but available for trade is possible. The trouble is that Ofek doesn't solve the issue or even really notice that before trade can drive evolution, it has to become an available behavior to be selected for. That's where evolutionary game theory comes in, searching for and with Axelrod, perhaps finding a behavioral

strategy of conditional cooperation that could be selected for and so lay the groundwork for trade. Haim Ofek, *Second Nature: Economic Origins of Human Evolution* (Cambridge: Cambridge University Press, 2001).

7. Lionel Robbins, *An Essay on the Nature and Significance of Economic Science* (London: Macmillan, 1932), 15.

8. Roger Meyerson, "Perspectives on Mechanism Design in Economic Theory," Nobel Prize lecture, December 8, 2007, https://www.nobelprize.org/uploads/2018/06/myerson_lecture.pdf.

9. Sebastian Krapohl, Václav Ocelík, and Dawid M. Walentek, "The Instability of Globalization: Applying Evolutionary Game Theory to Global Trade Cooperation," *Public Choice* 188 (2021): 31–51.

10. The top trading cycle has a feature that the med student-hospital match didn't allow: trading outcomes between willing participants. We'll ignore this wrinkle in what follows.

Bibliography

Arrow, K. "Uncertainty and the Welfare Economics of Medical Care." *American Economic Review* 53 (1963): 941–973.

Arrow, K., and G. Debreu. "Existence of an Equilibrium for a Competitive Economy." *Econometrica* 22, no. 3 (1954): 265–290.

Axelrod, R. *The Evolution of Cooperation*. New York: Basic, 1984.

Barro, R., and V. Grilli. *European Macroeconomics*. London: Macmillan, 1994.

Becker, G. *The Economic Approach to Human Behavior*. Chicago: University of Chicago Press, 1978.

Becker, G. *Human Capital*. Chicago: University of Chicago Press, 1995.

Becker, G. "Irrational Behavior and Economic Theory." *Journal of Political Economy* 70, no. 1 (1962): 1–13.

Bonnan, A., and S. Lopez. "Walmart and Local Economic Development." *Economic Development Quarterly* 26, no. 4 (2019): 285–297.

Cass, D. "Optimum Growth in an Aggregative Model of Capital Accumulation." *Review of Economic Studies* 32, no. 3 (1965): 233–240.

Christiano, L., M. Eichenbaum, and M. Trabandt. "On DSGE Models." *Journal of Economic Perspectives* 32, no. 3 (2018): 113–140.

Christiano, L., M. Eichenbaum, and M. Trabandt. "Understanding the Great Recession." *American Economic Journal: Macroeconomics* 7, no. 1 (2015): 110–167.

Duesenberry, J., E. Fromm, and L. Klein. *The Brookings Quarterly Econometric Model of the United States*. Chicago: Rand McNally, 1965.

Fehr, E., and U. Fischbacher. "Third-Party Punishment and Social Norms." *Evolution and Human Behavior* 25, no. 2 (2004): 63–87.

Frank, R. *The Darwinian Economy*. Princeton, NJ: Princeton University Press, 2011.

Friedman, D. *Price Theory.* La Jolla, CA: South-west Publishing, 1990.

Friedman, M. "The Methodology of Positive Economics." In *Essays in Positive Economics*, by M. Friedman, 3–43. Chicago: University of Chicago Press, 1953.

Friedman, M. *The Permanent Income Hypothesis: A Theory of the Consumption Function.* Princeton, NJ: Princeton University Press, 1957.

Friedman, M. "The Role of Monetary Policy." *American Economic Review* 58, no. 1 (1968): 1–17.

Hayek, F. *Law, Legislation and Liberty.* Vol. 3, *The Political Order of a Free People.* London: Routledge & Kegan Pau, 1979.

Hayek, F. "The Use of Knowledge in Society." *American Economic Review* 35, no. 4 (1945): 519–530.

Hobbes, T. *Leviathan.* London, 1651; Project Gutenberg, 2002. https://www.gutenberg.org/files/3207/3207-h/3207-h.htm.

Hobbes, T. *Leviathan*, edited by E. Curley. Indianapolis: Hackett, 1994.

Hume, D. *Enquiry Concerning Principles of Morals.* London: A. Millar, 1751.

Kaldor, N. "Capital Accumulation and Economic Growth." In *The Theory of Capital*, edited by F. A. Lutz and D. C. Hague, 177–222. London: Macmillan, 1961.

Keynes, J. M. "The General Theory of Employment." *Quarterly Journal of Economics* 51, no. 2 (1937): 209–223.

Keynes, J. M. *The General Theory of Employment, Interest, and Money.* London: Macmillan, 1936.

Keynes, J. M. *A Tract on Monetary Reform.* London: Macmillan, 1924.

Krapohl, S., V. Ocelík, and D. M. Walentek. "The Instability of Globalization: Applying Evolutionary Game Theory to Global Trade Cooperation." *Public Choice* 188 (2021): 31–51.

Lucas, R. "Macroeconomic Priorities." *American Economic Review* 93, no. 1 (2003): 1–14.

Lucas, R. "Nobel Lecture: Monetary Neutrality." *American Economic Review* 104, no. 4 (1996): 661–682.

Lucas, R., and T. Sargent. "After Keynesian Macroeconomics." *Federal Reserve Bank of Minneapolis Quarterly Review* 321 (Spring 1979): 1–16.

Mankiw, N. G. *Macroeconomics.* New York: Worth, 2010.

Meyerson, R. "Perspectives on Mechanism Design in Economic Theory." Nobel Prize lecture, December 8, 2007. https://www.nobelprize.org/uploads/2018/06/myerson_lecture.pdf.

Bibliography

Misak, C. *Frank Ramsey: A Sheer Excess of Powers*. New York: Oxford University Press, 2020.

Ofek, H. *Second Nature: Economic Origins of Human Evolution*. Cambridge: Cambridge University Press, 2001.

Olson, M. *The Logic of Collective Action*. Cambridge, MA: Harvard University Press, 1965.

Ostrom, E. *Governing the Commons: The Evolution of Institutions for Collective Action*. Cambridge: Cambridge University Press, 1990.

Phillips, A. W. "The Relation between Unemployment and the Rate of Change of Money Wage Rates in the United Kingdom, 1861–1957." *Economica* 25 (1958): 283–295.

Rabin, M. "Psychology and Economics." *Journal of Economic Literature* 36, no. 1 (1998): 11–46.

Ramsey, F. P. "A Mathematical Theory of Saving." *Economic Journal* 38, no. 152 (1928): 543–559.

Robbins, L. *An Essay on the Nature and Significance of Economic Science*. London: Macmillan, 1932.

Romer, D. *Advanced Macroeconomics*. 2nd ed. New York: McGraw-Hill, 2001.

Roth, A. E. "The Economics of Matching: Stability and Incentives." *Mathematics of Operations Research* 7, no 4 (1982): 617–628.

Roth, A. E., T. Sönmez, M. Ünver, and U. Utku. "Kidney Exchange." *Quarterly Journal of Economics* 119, no. 2 (2004): 457–488.

Scheidel, W. *The Great Leveler*. Princeton, NJ: Princeton University Press, 2017.

Schelling, T. "Hockey Helmets, Concealed Weapons and Daylight Savings." *Journal of Conflict Resolution* 17 (September 1973): 381–428.

Schelling, T. *The Strategy of Conflict*. Cambridge, MA: Harvard University Press, 1981.

Selten, R. "A Simple Model of Imperfect Competition, where 4 Are Few and 6 Are Many." *International Journal of Game Theory* 2 (1973): 141–201.

Shapley, L., and H. Scarf. "On Cores and Indivisibility." *Journal of Mathematical Economics* 1 (1974): 23–37.

Simon, H. A. *Models of Man: Social and Rational*. New York: Wiley, 1957.

Smith, A. *The Theory of Moral Sentiments*. London: Andrew Millar, 1759.

Smith, A. *The Wealth of Nations*. London: Strahan and Cadell, 1776.

Smith, J. M. *Evolution and the Theory of Games*. Cambridge: Cambridge University Press, 1982.

Solow, R. "The State of Macroeconomics." *Journal of Economic Perspectives* 22, no. 1 (2008): 243–246.

Stiglitz, J. "Where Modern Macroeconomics Went Wrong." *Oxford Review of Economic Policy* 34, no. 1–2 (2018): 70–106.

Thaler, R. H. "Behavioral Economics: Past, Present, and Future." *American Economic Review* 106, no. 7 (2016): 1577–1600.

Wilson, D. S., E. Ostrom, and M. E. Cox. "Generalizing the Core Design Principles for the Efficacy of Groups." *Journal of Economic Behavior & Organization* 90 (2013): S21–S32.

Woodford, M. "Optimal Interest Rate Smoothing." *Review of Economic Studies* 70, no. 4 (2003): 861–886.

Index

Advanced Macroeconomics (Romer), 62–63, 66, 115
Adverse selection, 216, 217
Agent-based models, 209
Aggregate demand curve, 118
Air pollution abatement equipment, as public good, 163
Algebra, elementary, 215
Allocative efficiency, 134, 156, 189
Amazon, 131–132
Animal spirits, 95, 101, 183
Antimissile defenses, 190
Arms race, 168, 190, 195, 212, 217
Arrow, Kenneth, 55, 126
Assuming the worst about people, 209
ATMs, 82
Auctions, 210–213
Axelrod, Robert, 203, 209, 217, 218

Banking, 124–125
Bank of England, 112
Barriers to entry, 2
Barro, Robert, 119
Becker, Gary, 29–33, 119, 154
Behavior, actual, 26
Behavioral economics, 21–27
Behavioral finance, 25
Behaviorism, 28–29
Billiard ball model of gas, 17–19, 27, 29, 121
Biology, 4, 96

Bohr's theory of atom, 27
Bonds, 78, 82, 83, 185
  stock, 186
Bosons, 6
Boyle's law, 16, 121
Budget constraint, 62, 65
Budget line, 30–32
Business cycle, 87, 101
Buyers and sellers in capital market, asymmetry between, 184. *See also* Labor market
Buyer's surplus, 139. *See also* Consumer's surplus

Calculation to coordinate plans, 199
Calculus, 59–71
Capital goods, 182, 183, 184, 213
  producers, smaller number of, 184, 189
Capitalism, 159
Capital stock, 195
Carrots and sticks, 214
Central banker's model of macroeconomy, 108–114
Central planning, 200
Chains of preference vs. cycles, 224
Choices, actual, 70
Chrysler Motors, 177–179
Civilization, 7, 166, 204
  unraveling, 198
Climate change, 165

Coalitions, 171, 188, 193. *See also* Collective action problem; Size (of group) problem
  of employees, 176, 177
  of employers, 148
  of market makers, 19
  monopolistic, 187
  monopsonistic, 188
Coase, Ronald, 205
Cold War, 190
Collective action problem, 149, 166, 175–176, 184, 187
Collusion, 166
Common-pool resources, 170–171, 172
Competitive market, 141
Computer simulations, 203, 218
Constant returns to scale, 135
Consumer's surplus, 142, 155, 157
Consumption, 194
  goods, 31, 64, 68–71, 72, 158, 163
  postponable, 165–167, 168, 172–173
  producers, large numbers of, 184
Cooperation, 202, 220
  in-group, 201
Corner-cutting, 208
Corruption, 214
Creative destruction, 157
Currency arbitrage, 190
Cycles of updating predictions and explanations, 5, 20, 27, 36, 108

Darwin, Charles, 201
Darwinian cultural selection, and spontaneous order, 199, 200, 205, 206, 207
Data, 4, 5, 6, 27–29, 46, 50, 51, 66, 67, 82, 99, 102, 104, 106, 110, 114, 117, 126, 129–131, 143, 170, 218, 220
  and modeling, 16–20, 95–96
  modeling by macroeconomists, 122–123, 128–131
  science as driven by, 3–5, 143, 170
  and stylized facts, 34–36, 85

Dead weight loss, 142, 146, 147, 157
Debreu, Gerard, 55, 126
Demand curve, 26, 43, 50, 57, 60, 82, 93, 100–101, 107, 119, 138, 140, 211. *See also* Slope of curve
  actual, 20, 35
  actual, long-run aggregate, 118
  downward sloping, 30, 32, 35, 49, 59, 69, 120
  in externalities, 155–157
  in monopoly, 141–145
  in monopsony, 145–147
Democratic governments, 195
Depressions, economic, 5, 88, 89, 92, 94, 96, 100, 101, 102, 191
Descent of Man, The (Darwin), 201
"Design space," 205
Diamond, Peter, 150
Differed acceptance algorithms, 223
Differential calculus, 59, 60, 68, 69, 70
  equations, 60, 70
Diminishing marginal rate of substitution (DMRS), 51–52, 140
Diminishing marginal utility (DMU), 43, 46, 51–52, 59, 139–141
Discipline of optimizing agents and market clearing, 97, 103, 124, 131, 149. *See also* Lucas, Robert; Sargent, Thomas
Discounting, 65. *See also* Hyperbolic discounting
Division of labor, 181, 182, 185, 195, 205, 206
DNA sequences, 96
Dodd-Frank Act (2010), 127, 191
Dominant strategy, 211, 224
Donors, kidney, 223–225
Double coincidence of wants, 74–77, 79
  intertemporal, 82, 183, 185, 205
Dynamic rationality, 106
Dynamic Stochastic General Equilibrium (DSGE) models, 108–131

Index

e, mathematical constant, 63
Econometrics, 116
Economically unproductive activities, 195
Economic Approach to Human Behavior, The (Becker), 32
Economic imperialism, 32
Economic nationalism, 220
Economics
　definition of, 49, 208
　as insulated from empirical refutation, 58
　most important job of, 198
　natural state of, 114–120
Economic theory as descriptive vs. normative, 22, 49, 57, 58, 129. *See also* Macroeconomic theory: as moral philosophy
Economists
　advice of, 2
　defunct, 1
　explanations "feel right" to, 5
Edgeworth, Francis Ysidro, 41, 59
　box, 52–54, 56, 60, 62, 79, 89
Efficiency, 40, 48
Efficiency wages, 124
Egalitarianism, 213, 215
Einstein, Albert, 69
Electromagnetic spectrum auctions, 211
Empirical data, 20, 67, 69
Empirical evidence, 36, 99, 128, 129
Empirical facts, 33, 49
Empirical models, 16, 29
Empirical research, 26
Empirical science, 4, 7, 10, 13, 24, 48, 95, 99, 102, 126, 128, 130
　realistic explanations vs. idealized models, 26
Employer search, 152
Endowment effect, 43
Entrepreneurship, 137, 149, 157, 167
Environmental damage, 156
Environmental economics, 155, 157

Equality, 48
Equations, 2, 99
Equity vs. efficiency, 40, 45–49
European Central Bank, 112
Evolutionarily stable strategies, 202
Evolutionary anthropology, 201
Evolutionary game theory, 201, 206, 220
Evolutionary just-so stories about money, 75–77
Evolution of Cooperation, The (Axelrod), 203
Exchange, 13
Exchange economy, 73
　real prices in, 74
Exogenous forces, 88, 125, 137
Exogenous shock, 113, 124
　to demand, 116
Exogenous variables, 82, 109
Expected utility, 91
Experimental game theory, 185, 188, 206, 209
Experimental research program, 36
Explanation and prediction, 4
　by models, 16–20, 95
Explanatory, predictive role for mathematics in economic theory, 70
Explanatory target, of economics, 33
Exploitation of labor, 146
Externalities, 154–157

Fairness norm, 24
Federal Reserve Bank, 112, 129, 191
Fermat, Pierre, 38
Fermions, 6
Fiat currency, 205
Financial institutions, role of, 185
Financial market, 90, 159, 178, 181–195
　ideal role of, 184–187
　impact on perfect competition, 192
　inevitable market failure of, 187–191
　kinds of agents in, 182–187
　regulation of, 188, 191

Financial panics, 88
Firms, 54, 62
First theorem of welfare economics, 14, 33, 39, 47, 74, 94, 100, 103, 108, 114, 128, 133–135, 137, 158
  assumptions of, 55–57
  proof of, 49–58
  role in macroeconomics, 62–64, 69, 115, 122, 126
Fisher, I., 110
Focal point, 179
Food chain, 201
Ford, Henry, 166
Ford Motor Company, 177–179
Frank, Robert, 176
Free market, 47, 55, 114, 123, 128, 133, 179, 200, 206
Free rider, 165, 167, 170, 171, 174, 176, 177, 179, 209
Frictional unemployment, 154
Frictions, 123–125, 127, 152–153, 179
Friedman, Milton, 28–29, 33, 115, 120, 123–125, 127, 154, 198
  on Phillips curve, 104–106, 107, 109, 111, 118, 150, 153
Full employment, 106
Full information, 189, 210
Full value bids at auction, 211–212
Future consumption, 63
Futures market, 134

Gamblers, 22, 183
Game theory, 9, 10, 24, 89, 133, 143, 149, 154, 158
  in auction design, 209–213
  evolutionary, 201–206
  and financial markets, 181–195
  in institution design, 206–209, 212, 215, 216, 217, 218, 220–225
  introduced, 159–179
  as lifesaving tool in health care, 221
  principal-agent problem, 213–218

Gaming the system, 133, 191, 195, 209, 210
Gases, models of, 16–20
  behavior of, 121
  billiard ball, 17–19, 27, 29, 121
  real, 18
Gas laws, 121. *See also* Ideal gas law
Gas stations, 145
Gathering "for merriment and diversion," 148, 162, 172
Gay-Lussac's law, 121
General equilibrium, 48, 52, 74, 84–85, 101, 110, 113, 124, 131, 150, 152, 158, 179. *See also* Pareto optimal general equilibrium
  existence of, proved, 54–57
  level of output, 110, 126
  in macroeconomy, 60–66, 115–120
General Motors, 177–179
General Theory of Employment, Interest and Money, The (Keynes), 89, 96, 100
General theory of relativity, 69
Generous tit for tat strategy, 218
Genetics, 96, 123
Glass-Steagall Act (1933), 191
Gold and silver, 86, 87. *See also* Money
Good ideas, 166–169
Government, 1, 170
  deficits, 103
Graphs, 99, 156
Grassroots organizations, 171
Great Depression, 5, 89, 94, 96, 100, 102, 191. *See also* Depressions, economic
Growth theory, 62

Hamilton, William, 203
Happiness, maximization of, 41
Harmless assumptions, 56
Hayek, Friedrich, 199
Hegemony of theory, 5
Herrnstein, Richard, 22

Index

Hobbes, Thomas, 10, 165, 170
Hobbes's foole, 11, 197–199, 201, 206, 207, 208, 209, 220, 224
Hockey helmets, collective action problem, 175
Hodgkin–Huxley model of neuron, 27
Home ownership, 177
Homo economicus, 7, 8, 11, 14, 16, 37, 198
Homo sapiens, 200
Hospital administrations, 221
Households' lifetime utility, 62–65, 114–116
   consumption vs. savings, 115
   infinite time horizon, 66
Human behavior, 14
Human capital, 33
Hume, David, 11, 85, 87, 197–198
Hume's knave, 11, 197, 198, 201, 206, 207, 208, 209, 213, 220, 224
Hurwicz, Leonid, 206, 208
Hyperbolic discounting, 23, 25, 78, 183

Ideal gas law, 16, 17–20, 27, 87, 95, 121
Idealizations, 158
Idealized models, 70, 102
Impulse-response curves, 129–130
Incentive compatibility
   algorithm, 222
   mechanisms, 206–225
   need for, 220–225
Income effects, 34–35
Income tax, 209
Indifference curves, 49–54, 69
Industrial Revolution, 157
Inequality
   forever increasing, 192
   produced by capitalism, 213
   in productivity, 192
Inflation, 105, 111, 117
Inflationary supply shock, 119
Information, complete, 189, 210
Information storage, 199

Initial endowment, 53
In-person work, 148
Insatiability, 14
Institutions, 4, 7, 11
   design, 10, 128, 167, 195, 197–225
Integral equations, 64
Intellectual property, 166–169
Intentional design, 75, 77, 200
   by humans, 205
Interest rate, 7, 90, 100, 102, 105, 119, 122, 125, 184, 186, 192
   and central bank, 127–129
   expected vs. actual, 112
   in macroeconomic model, 108–117
   and money balances, 80–83
   natural rate of, 116
   real vs. nominal, 110–111
International trade, 217–220
Intractable problems, 217
Introspection, 27, 41–42, 43, 46, 51, 67
Invisible hand, 13, 14, 40, 46, 48, 57, 122, 128, 134, 181, 189
"Irrational Behavior and Economic Theory" (Becker), 29–33, 120
   impulsive and inertial consumers, 30–32
Iterated prisoner's dilemma, 204, 218–219. *See also* Prisoner's dilemma

Job searching, 179
Justice, 40

Kahneman, Daniel, 22
Kandel, Eric, 122
Kennedy administration, 104
Keynes, John Maynard, 1, 183
   as behavioral economist, 95, 101
Keynesian economics, 8
   models, 101, 102, 122
   planning, 95
   policy, 118
   revolution, 6, 99

Keynes's flat ocean, 89, 106, 118, 122
Kidney transplantation, 223–225
  knavish donors, 224
Koopmans, Tjalling, 114
Krapohl, Sebastian, 217–218
Kremer, Michael, 169
Krugman, Paul, 71

Labor, sellers of, 133
Labor market, 9, 88, 101, 113, 145, 159. *See also* Matching problem
  collective action problems of, 171–179
  collides with economic theory, 149–154
  as monopsonistic, 147–149
Labor vs. leisure, 64
Lagrangian, 61–62, 115
  in Romer's equations, 66, 70
Lavoisier, Antoine, 4
Law of one price, 186
Laws of supply and demand, 32
Leaving money on the table, 25
Lens, in Edgeworth box, 52–53
Leviathan (Hobbes), 166, 197
Lifetime consumption, 106
Limited information, 107. *See also* Information storage
Liquidity, 83
Locke, John, 5, 10
Logical positivism, 28–29
"Long moderation," 123
Long-run equilibrium, 113, 117, 118, 119, 145, 158. *See also* General equilibrium
  vs. short-run, 104, 105, 106, 108, 114, 115, 116, 127, 136, 137, 150, 195
  time periods, 84, 88–89, 128, 136, 153, 158, 172
Long-term investment, 194
Loss and risk aversion, 22, 24, 26
Lucas, Robert, 97, 102, 122, 149

Macroeconomic models, 99, 108
Macroeconomics, 7
  history of, 99–107
Macroeconomic theory, 6, 8, 73–97
  as moral philosophy, 41, 46, 48–49, 112–115, 126–128
Macro variables, 100
Malthus, Thomas, 191
Management, bad and good, 214, 215
Managers, 213–217
  moral hazard constraint of, 213
Mandeville, Bernard, 181–182
Mankiw, Gregory, 109, 113
Marginal cost curve, 138, 141
Marginalist economic theory, 42–43, 96
Marginal product of labor, 146, 147, 151
Marginal propensity to save, 106
Marginal revenue curve, 143
Marginal utility, 59. *See also* Diminishing marginal utility
Market clearing, 57, 87, 97, 100, 103, 107–109, 110, 111, 122, 145, 148, 150, 153, 154
  equilibrium, 74, 84, 88, 89, 101, 124, 126, 131, 137, 178 (*see also* Discipline of optimizing agents and market clearing; General equilibrium)
Market competition, as morally ideal, 48
Market failure, 133–158, 165, 181, 185, 189
Market maker. *See* Middlemen
Market power, 138, 193, 194
Market rationality, 29–33, 37, 120
"Markets clear, agents optimize," mantra of new classical macroeconomics, 97, 103. *See also* Discipline of optimizing agents and market clearing
Marx, Karl, 134
Matching problem
  in employment, 150–154
  in internship selection, 221–222

Math(s), 2, 8, 59–71
Mathematical expectation, 91
Mathematical game theory, 189, 201, 210, 224
Mathematical models, 13
  proof, 54, 101
  tractability of, 70
Mathematical Psychics (Edgeworth), 41, 44, 52
Mathematizing utility, 59–60
Maximizing revenue, 143
Maximizing total utility, 46
Measurable units, 42. See also *util*
Mechanism design, 11, 197–225
Mendel, Gregor, 96, 122
Mendeleev, Dmitry, 4
Mendelian genetics, 27
"Methodology of Positive Economics, The" (Friedman), 28, 58, 69–70, 85
Meyerson, Roger, 213
Microeconomic theory, 6, 7
Microfoundations, 103, 108, 114–117, 119, 125, 128, 130, 131
  necessity of, 120–123
Micro-modeling in natural science, 107–108, 123
Middlemen, 181, 187, 188
Mobile-telephone companies, 211
Models working with data, 19. *See also* Cycles of updating predictions and explanations
Money, 8, 22, 43, 59–71, 134, 182, 205. *See also* Store of value
  as accounting tool, 77, 183
  causal role of, 77, 83, 87, 88, 93, 94, 113–114, 183
  as a claim on future consumption, 182–183
  five features of, 74–75
  illusion, 83, 84, 87, 88, 92, 103, 106, 117
  inertness of, 83
  laundering, 190
  neutral, 183 (*see also* Neutrality of money)
  precautionary role of, 91, 92
  prices, 88
  shiny bits of metal, 75, 79, 86, 205 (*see also* Gold and silver)
  speculative role of, 89
  supply, 100
Monopolistic, monopsonistic coalitions, 153–156, 190
Monopolist's strategic action problem, 211
Monopoly, 138–145, 155, 199
  can't last, 171
  extent of, 144–145
  harms of, 138–144
  model of, 142
  nonexistence of, 143–145, 171–172
Monopoly (board game) 208
Monopsonists vs. monopolists, games between, 177–179
Monopsony, 139, 169, 170, 173, 199
  in labor markets, 147–149, 150, 153–155, 171, 177, 179, 190
Moral hazard, 215, 216
Morality, 46
Moral philosophy, 48, 198. *See also* Macroeconomic theory: as moral philosophy
Moral preferability, 58
Mortenson, Dale, 150
Multigenerational wealth transfers, 193

Natural level of output, 111, 117, 126, 128
Natural rate of interest, 116
Natural rate of unemployment, 106, 109, 111, 150, 153
Natural state of economy, 111, 114–120
Negative externalities, 155
Neoclassical economics, 96

Neutrality of money, 83, 84, 85, 96. *See also* Money
New classical economic theory, 6, 9, 73–97, 145, 153
and subprime mortgage recession, 123–126
New Keynesians, 124, 150
Newton, Isaac, 4
Newton's laws, 17, 18
Nobel Prizes, 3, 7, 11, 21, 25, 28, 29, 33, 55, 69, 91, 96, 102, 125, 150, 169, 170, 178, 196, 199, 205, 206, 210, 213, 223
Nominal prices, 119
Nominal rate of interest, 127
Non-acceleration inflation rate of unemployment (NAIRU), 153
Non-excludable consumption, 165. *See also* Public good
Nonoptimal equilibrium, 100. *See also* Keynesian economics: models
Non-rivalrous consumption, 165. *See also* Public good
Normative force of RCT, 22. *See also* Moral philosophy
Normative use of economic theory, 37
Norm-following, preference for, 24

Ocelik, Vaclav, 217–218
Olson, Mansur, 170, 184
Omniscience, as requirement for RCT, 14
Online job postings, 149
On the Origin of Species (Darwin), 8
Operating rooms, 224
Opportunity cost, 80
Optimal growth, 193
Optimality of general equilibrium, 150. *See also* Pareto optimal general equilibrium
Optimum cash balance, 78–80, 83
Ordinal preferences, 45
utility, 46

Ostrom, Elinor, 170–171, 172, 184, 187
conditions for coalition stability, 171, 173–175
and labor market monopsony, 171–175
Oversupply of public goods, 170

Pareto, Vilfredo, 45, 47, 48, 66
Pareto improvement, 53
Pareto optimal general equilibrium, 47, 48, 52, 53, 57, 62, 63, 87, 88, 89, 95, 96, 100, 103, 106, 115, 127, 128, 136, 137, 145, 207, 210
Pareto optimum, 224
Partial differential, 60, 61
Participation constraint, 213, 216
Patents, as second best, 167–169
Pavlov, Ivan, 122
Payoff matrix, 160, 218, 219
"People of the same trade," 148. *See also* Gathering "for merriment and diversion"
Perfect competition, 114, 134–135, 157, 167, 198, 217
in financial markets, 185
incentive compatibility of, 207
wages, 147
Permanent income hypothesis, 94, 102, 106, 115, 133
Perverse incentives, 207
Pharmaceutical patents, 169
Phillips, William, 93
Phillips curve, 93, 94, 101, 104, 110, 111, 117, 150, 153
long-run vs. short-run, 104–106
Physicians, 220–223
Physics, 1, 3, 19, 26, 59, 69, 96, 121
Pissarides, Christopher, 150
Plato, 8, 197
Politicians, 2
Pollution, 154–155
Poor, the, 191

Index

Positive externalities, 155, 165
Positive monotonic transformations of utility functions, 68–69
Postcards from *Advanced Macroeconomics* (Romer), 63, 66
Postponable consumption, 182, 185. *See also* Consumption: goods, Money
  rationed by interest rate, 187
  of wealthy, 194
Precautionary balances, 89, 92
Predictive accuracy, 5, 18
Predictive power, 4, 5, 16, 21–23, 25, 26, 29, 185
  of models, 51, 63, 84, 107–109, 110, 121–122
Preference ranking, 20, 28, 177, 179
  in games, 160, 161, 164
Prescriptive models, 126
Price movements, actual, 36
Prices, 20. *See also* Real, relative, and money prices
Price setters, 187, 159, 194
Price system, evolutionary origin of, 200. *See also* Spontaneous order
Price takers, 56, 138, 143, 147, 151, 176, 199, 207
Priestly, Joseph, 4
Principal-agent problem, 212–217
Principia Mathematica (Newton), 8
Prisoner's dilemma (PD), 159, 160–164, 178, 201. *See also* Iterated prisoner's dilemma
  iterated, 204, 218–219
  and supply of public goods, 159–164
  trading as, 202
Privatizing, 208
Probability, 21, 90–91
  distributions, 107
Producers, 54
Productivity of economy, 185
Profit, real, 133–158. *See also* Market failure; Rents and rent seeking

Proof. *See also* First theorem of welfare economics
  of market optimality, 134
  turning Smith's conjecture into, 49–57, 210
Psychologism, 28
Psychology, 149, 153–154. *See also* Behavioral economics
  causes in, 44
  of choices, 28
  explanations, 44
  facts, 68
  models, 27
  theory, 26, 34
Psychophysics, 44
Public good
  defined, 162–163, 167
  provision, 169–171
Punnett squares, 27

Quantitative vs. stylized facts, 34
Quantity theory of money, 85–89
  and gas law, 86

Ramsey, Frank, 114
Ramsey-Koopmans-Cass model, 115, 116, 119
Random shocks, 110, 111. *See also* Exogenous shock
Rational agents, 13–37, 88
  bargainers, 53
Rational but fallible expectations, 98, 100, 103, 104, 107, 116, 120
Rational choice theory (RCT), 7, 13–37, 38, 40, 41, 44, 50, 64–65, 72, 73, 79–81, 84, 85, 86, 87–89, 94, 97, 106, 107, 122, 151
  explanation of stylized facts, 36
  as ideal model, 15–17, 18–21
  indispensability of, to economics, 25, 39–58
  normative, 25
  not a psychological theory, 27–28, 33
  stated, 14

Rational egoism, 24
Rational expectations, 104, 107, 150
 fallible, 113
RCT-rational bots, 209
RCT-rational "fooles," 188, 189
Real, relative, and money prices, 73–74, 75, 83
Real economy, 8, 30, 55, 57, 71, 73, 82, 99, 100, 103, 126, 127, 144, 145, 157, 166, 171, 179, 187, 189, 195, 198, 199, 212
 money in, 82–87, 92, 93, 96, 124, 182
 profit in, 134–137, 167
Real estate market, 176
Realism of models, 29, 39
 increasing, 18, 24
Real preferences, 207
Real price vs. money price changes, 106
Recessions, defined, 119
Redistribution, 194
Relative reward problem, 176
Remote working, 149
Rental market, 176
Rents and rent seeking, 136, 155, 166, 187, 188, 189, 190, 208, 212, 217
 and monopoly and monopsony, 138, 142, 145, 146, 147
 and the wealthy, 192–195, 199
Republic, The (Plato), 197
Revealed preferences, 28, 45, 46, 107
Rich will always be with us, 191–195
Right to work (at lower wages) laws, 177, 178, 179
Rip Van Winkle effect, 99–103
Risk, as a convention, 91
 and well-behaved probabilities, 90
Robbins, Lionel, 49, 62, 208
Robinson Crusoe economy, 114
Roll-over crisis, 124
Romer, David, 62, 65, 66, 67, 115
Rotating crops, 167
Roth, Alvin, 222–225

Rousseau, Jean Jacques, 10, 177
Rule of law, as a public good, 165–166

Salary cap, in professional sports, 176
Samuelson, Paul, 28
Sargent, Thomas, 97, 109, 122, 149
Satisficing, 25
Savings, 194
Scarf, Herbert, 222
Schelling, Thomas, 175, 178
School choice, 224–225
Schrodinger's wave equation, 63
Science, 2–3, 4–8, 9, 10, 13, 32, 36, 44–45, 69–70, 71, 76, 96, 102, 108, 128, 130, 199, 206, 228, 232, 235
 behavioral, 6, 26, 33, 158, 201
 models in, 16, 17–20, 21, 26, 27, 57–58, 86, 108, 122–123, 126
 natural, 33
 research programs in, 27, 96
 as value neutral, 48, 49, 57
Search model of labor market. *See* Matching problem
Second-best solution, 167–169, 191
Second price auctions, 211–212
Second World War, 200
Selfish vs. unselfish strategies, 24
Self-liquidating, monopoly and monopsony as, 150
Sellers and buyers in capital market, asymmetry between, 184. *See also* Labor market
Setting prices, monopolist and monopsonist problem of, 168
Shapely, Lloyd, 222
Shiny bits of metal, 75, 79, 86, 205. *See also* Gold and silver; Money
Shock, financial, 125. *See also* Exogenous shock
"Shut up and calculate," 69
Simon, Herbert, 25
Size (of group) problem, 149, 166, 170
Skinner, B. F., 45

Index

Slope of curve, 20, 25
   demand curve, 20, 30, 32, 35, 80, 82, 141, 143, 147, 211
Smith, Adam, 6, 8, 9, 13–14, 39–58, 86, 99, 113, 123, 128, 134, 138, 148, 158, 162, 172, 181, 182, 191, 201, 202, 206, 225
   conjecture of, 13, 14, 38–39, 40–41, 46–47, 48, 49, 50, 52, 54, 57, 134, 158
Smith, John Maynard, 201
Social contract, 10
Socialism, 213, 217
Social psychologists, 21
"Social Uses of Knowledge, The" (Hayek), 199
Sociopaths, 11, 209
Solow, Robert, 123
"Sort through the wreckage," 97, 103, 122
Speculative balances, 89, 92
Spontaneous order, 199, 200, 205, 206, 207
Squinting, 55, 157, 193
Stable preferences, 32
Stagflation, 95, 123
Stag hunts, 177, 178, 190
Stalin, Josef, 214
Statistical tools, 3
Sticky prices, 124, 125, 150
Sticky wages. *See* Wages, sticky
Stiglitz, Joseph, 125, 126
Stochastic shocks, 117, 129
Store of value, 77–78. *See also* Money: as a claim on future consumption
Strategic Defense Initiative, 190
Strategic interaction problem, 9, 143, 147, 158, 168. *See also* Game theory
Street lighting, as public good, 163
Stylized facts, 29–32, 54, 61–63, 72, 73, 75, 83, 84, 85, 89, 108, 111, 123, 125, 126, 131, 132, 135, 136, 138, 139, 140
   and calculus, 69–70

Subjective probabilities, 91
Subprime mortgage recession, 123–126
Substitutes, 36
Substitution effects, 34–35
Suckers, 163
Supply curve, 143
   long-run aggregate, 118
Surpluses and gluts, 100
Surplus production of consumption goods, 184, 189
Swedish State Bank, 7

Tariff barriers, 218
Tax authorities, 76
Tax rates, 195
Technological change, 82, 193
Textbooks, as foundational, 3, 5, 138
Theory-driven discipline, economics as, 4
Theory of consumer behavior, 120
Theory of firm, 120
Theory of the Moral Sentiments, The (Smith), 40
Thought experiments, 55, 65, 67, 109, 116, 117, 123
Tit-for-tat strategy, 218, 203, 220
"Too big to fail," 190
Top trading algorithm, 223
Top trading cycles, 222, 224, 225
Tract on Monetary Reform, A (Keynes), 89
Trade secrets, 193
Trade unions, 178
Trading, as a prisoner's dilemma, 202
Traffic laws, as public good, 165
Transaction costs, 2, 78, 176, 188
Transitivity of preferences, 14
True preferences, 224
Tversky, Amos, 22

Uber, 148
Uncertainty, 89, 92, 101, 183
Unconditional altruists, 224
   cooperation among, as strategy, 218

Unconditional defection strategy, 218
Unemployment, 92–95, 100, 102, 104–106, 108–109, 111, 113, 118, 124, 129, 149–154, 171
Unilateral free trade, 218–220
United Automobile Workers, 177–179
United States, 190
  auto industry, 177–179
  depressions and financial panics, 88
  government, 211
Units of time, to measure long and short terms, 84
USSR, 190, 200
util, 42
  as measurable, 59
  unit of utility, 43
Utilitarianism, 41, 42, 46
Utility, 41–52, 54, 59–71, 97, 106, 114, 115, 119, 126, 140
  equations, 63, 66–67, 70
  expected, 91
  functions, 70
  lost in monopoly and monopsony, 124–128, 141–142, 143, 147
  as mathematical convenience, 68–69
  no such thing as, 42–44

Value function, 22, 24
Van der Waals equation, 19, 121
Velocity
  of money, 75–77, 79, 83
  of transactions, 88 (*see also* Quantity theory of money)
Vickery, William, 211
Von Neumann, John, 90, 91

Wage inequality, 192
Wage rates, 147
Wages, sticky, 92–93, 94, 125, 127, 150
Waiting lists, for transplants, 223–224
Walentek, Dawid, 217–218
Walmart, 131–132
Watson, James, 122

Wealth of Nations, The (Smith), 8, 12, 34, 87, 131, 162, 172
Welfare economics. *See* First theorem of welfare economics
Welfare maximization, 47
Western and non-Western cultures, 23
Wilson government (UK), 104
Worker search, 152
World Trade Organization, 190, 217–218, 220